ALWAYS FAITHFUL

ALWAYS FAITHFUL

A Story of the War in Afghanistan,
the Fall of Kabul, and the Unshakable Bond
Between a Marine and an Interpreter

MAJOR TOM SCHUEMAN
AND ZAINULLAH ZAKI WITH
RUSSELL WORTH PARKER

wm
WILLIAM MORROW
An Imprint of HarperCollinsPublishers

Photographs in the insert are courtesy of the authors, with the exception of page 8, top left, courtesy of Gina Pecho.

HarperCollins books may be purchased for educational, business, or sales promotional use. For information, please email the Special Markets Department at SPsales@harpercollins.com.

FIRST EDITION

Designed by Bonni Leon-Berman

Library of Congress Cataloging-in-Publication Data has been applied for.

ISBN 978-0-06-326061-0

22 23 24 25 26   LSC   10 9 8 7 6 5 4 3 2 1

TOM and ZAK

To God, Country, Corps, and the young men who walk point

دالی:

خدای، هیواد او هغه امنيتي ځواکونو ته چې د جگړې په لومړۍ کرښه کې دنده اجرأکوي.

CONTENTS

PROLOGUE

Countdown: Spring 2020

ZAK

I STILL DREAMT of Sangin in those days. I'd see bullets dancing across the surface of an irrigation canal as Marines hugged the side to escape them. I'd smell ammonia and smoke and hear the screams of a wounded man. And then, I'd wake suddenly to find my wife in bed next to me and our children asleep in our home.

It was 2:34 A.M. on a Tuesday when I traded the machine gun fire in my dreams for the insistent tones of my cell phone. Awaking from a Sangin dream was like surfacing from deep water and suddenly reaching air, my heart would race, my breaths would come as if torn from my lungs. I needed time to remind myself where I was, to center myself in the rhythms of my wife's gentle breaths, in the sounds my children make in their sleep. But on this night, as the phone rang, I was confused, still half asleep, as I looked at the number on the screen. I did not recognize it, but given the hour, I assumed it was an emergency, "*Salaam Aleikum.*"

The voice on the other end was hoarse. "Is this Zainullah?"

"*Bale.*"

"We are coming for you, infidel. Your American masters are running from Afghanistan. Turn yourself in and beg for mercy or the Taliban will take your hea—"

I pushed END on the phone before he could say more. The room holding my family suddenly seemed to close in on me. My stomach felt as if it were being squeezed by a fist. There would be no more sleep that night. I checked each of my children. Deewa, my wife, stirred and was still. I got up to make sure the gate of our home was locked. I turned at the door to look at my wife and children, sleeping peacefully. They did not know what I knew. If I could help it, they never would.

The moonlight illuminated the mountains as I stepped upon the terrace of our home. The stones cold under my feet, I turned my eyes to the stars shining back at me, so beautiful and so uncaring. I checked the gate, again, and was relieved to see it still locked. My family was safe for the moment, but I wondered how much time we had left.

This wasn't the first such call I had gotten—it wouldn't be the last. It had been ten years since I answered the call of my nation and went to help the Americans as a translator in Sangin District of Helmand Province. I spent months at war alongside American Marines and the Afghan National Army. We walked mile after mile in a place where the ground itself seemed to be trying to reach up to drag us down with claws of flame and steel. After decades of war, I had wanted to help build an Afghanistan where we could have the things the Americans talked about: freedom, democracy, and security. I wanted our country to guarantee basic human rights for our citizens. I wanted my daughters to go to school with my son and have a say in their own future.

Since 2016 though, it had been clear my dreams for Afghanistan were only that. I had decided back then, that if I could not raise my children in a safe, free Afghanistan where they could choose their own paths in life, I would apply for the Special Immigrant Visa (SIV) offered to battlefield translators and Afghans who worked directly for the United States government. I had wanted to build a life in my country, but if that was impossible, we would seek our dreams in America as so many have done. I began the SIV process in 2016, but when the bureaucracy proved too much for me alone, I reached out to Major Tom Schueman, the man who led the Marines I worked with in Sangin, in hopes he could help. He immediately went to work to help me. But by 2020, even with both of us, the system was still too complicated.

Making things worse, time was running out for me and my family. In February 2020, the Americans and the Taliban signed a deal in Doha, Qatar, that would have all Americans gone from Afghanistan by May 2021. I knew that I had attached my future and that of my family to the hopes the Americans brought in 2001, but those hopes would leave with them. The Americans were leaving for good, and I feared what lay ahead.

As a former interpreter, I had long known I was a target and I had accepted the risks. Other interpreters and members of the Afghan National Security Forces had already been killed by the Taliban, who posted "Night Letters" in public places threatening death to people who worked with the Americans or the Afghan government. Sometimes they named names. Sometimes they were left on a specific person's doorstep or taped to their door. The phone calls threatening me had started back in 2018. Now, standing in the courtyard of my family's home in the early hours of a morning in spring 2020, I could still hear the hoarse voice and the threat it had delivered only moments before.

As I turned my eyes back to the stars, my phone, still silenced to allow my family to sleep, buzzed in my hand. I looked down. A text message had arrived.

"You cannot escape us, Zainullah. The justice of Allah will be swift. Your day is coming."

TOM

AS WITH MOST things in the United States Marine Corps, there is a cadence to the ceremonies by which we honor Marines who make the ultimate sacrifice. At a military funeral, the family and friends sit waiting before the open grave. From the rear of a hearse, a pallbearer detail in Marine dress blues draws forth a flag-draped casket. A Marine Staff Noncommissioned Officer calls out, *"Present arms!"* Uniformed service members snap a salute as the pallbearers carry a brother or sister to the grave feet first, the flag positioned such that the blue field of stars is at the head of the casket and over the left shoulder of the Marine within. A rifle firing detail's weapons, stacked at the front right side of the gathering, await the Marines serving as pallbearers, or "body bearers" in the blunt nomenclature of the Corps. A bugler stands opposite. Silence lays heavy upon the shoulders of the living as the pallbearers quietly execute the commands required to place the casket upon the carriage in front of the grieving family. On a muted command, the pallbearers execute facing movements and march forth, merging into a single column as they retrieve their rifles and stand at parade rest for the duration of the service.

The conduct of the service varies by religion and individual family desires, but at the conclusion, the senior member of the funeral detail moves to the head of the casket for a final hand salute. He or she is joined by every uniformed member present as the rifle detail, seven Marines strong, fires three times for a total of twenty-one shots before moving the rifles to a position known as "present arms," a salute with a rifle. Marines in the crowd try not to flinch at the rifle fire. Their success often varies by the regularity and severity of their exposure to enemy fire. The bugler plays "Taps" as the echoes of rifle shots fade in the distance.

As the haunting, lonely, notes of the bugle fade into silence, the rifle firing detail stacks their weapons again and, in two ranks, marches back to the casket to fold the flag covering it. The senior member slowly lowers his salute. Marines and other servicemembers in the crowd follow suit, often accompanied by Marines now civilians. The salute seems discordant from men with long hair and beards, years of civilian life often gathered around their middles. But even with service sometimes decades past, the impulse to render honors is simply too deeply ingrained for men and women reborn as United States Marines at Parris Island, South Carolina; San Diego, California; and Quantico, Virginia.

Once the flag is folded, the senior pallbearer executes a right face and places it into the hands of the Marine designated to present it to the family. The movements take place in slow motion, a nod to the solemnity of the ceremony. The Marine carries the flag to the recipient, often a wife or mother, and says, "Ma'am, this flag is presented on behalf of a grateful nation as an expression of appreciation for the honorable and faithful service rendered by your loved one." Taking a moment to try and comfort someone often far beyond it, he or she returns to the head of the casket. It is that sameness, the predictability of the ceremony, that allows Marines to deal with the drumbeat consistency of funerals during wartime. It is that same predictability that is so shocking to the senses when those wartime experiences pull Marines long at peace into the darkness years later.

Corporal Justin McLoud was a standout high school baseball player who turned down college scholarships for life as a Marine rifleman. He was as superb a Marine as he had been an athlete. He had thrived during a 2008 tour in Fallujah, Iraq, a place synonymous with savagery. By 2010, Justin was a husband and new father beginning to consider the shape of life after the Marine Corps. But when it came time to decide whether to go to Afghanistan with me or leave for civilian life, he discussed it with his wife, then looked me in the eye and said, "You're my family, too. I'll stay."

In April 2020, almost ten years after leaving both his legs and an arm in Afghanistan, Corporal Justin McLoud died of an overdose of pain medication. As I watched the ceremony memorializing his thirty-two years on earth, surrounded by men who knew him at his best, I thought back to his worst moment. As he lay in the Afghan dirt, the ammonia stench of homemade explosive in the air around us, other Marines and I tried to help our Navy corpsman hold what remained of him together and begged him to stay alive for his infant son. He couldn't, but we would not know that for ten more years. Not all casualties actually die on the battlefield. Some are dead for years before we actually bury them.

As I finished eulogizing Justin McLoud at a Veterans of Foreign Wars post not far from the veteran's cemetery in which he now rests, I gazed across those in attendance to his young son and thought of the debts born of war. I could not help but think of all the men to whom I owed so much. My fellow infantrymen figured heavily in the calculus, of course, too many of whom had fallen to the hidden wounds of war in the years since Sangin. But on the day we buried Justin McLoud, as I looked at the faces of Marines I knew as teenaged heroes, now men entering their thirties, I was also thinking about someone who was absent, not because of death but because of distance: our Afghan interpreter, Zainullah "Zak" Zaki.

I hadn't heard from Zak for years until he'd reached out to me in 2016 on Facebook, seemingly out of the blue. He had been trying unsuccessfully to secure SIV visas for him and his family under a U.S. Department of State program for Afghan translators who had worked for the U.S. military. That Facebook message had been four years earlier, and ever since, I'd been trying, and failing, to help him navigate the byzantine red tape of the SIV process. But I would keep pushing till one or both of us was dead. I owed him everything I could give. America had made a promise to him. He was as much a brother as the Marines arrayed about me at the funeral.

Ever since Zak had first reached out to me, I'd known that there was an urgency. Zak was not one to leave Afghanistan without cause—he'd always believed in his country, in its prospects and in its people. But he and I both saw what was coming; now, with the May 2021 date for the U.S. withdrawal agreed upon, the Taliban threats he'd been receiving had started coming more frequently. And they only had to get lucky once.

I had attended enough funerals. I'd lost enough friends to a war that somehow was not done taking victims. I had helped save Corporal McLoud's life in 2010, only to lose him ten years later. I could not lose Zak, too. The clock was ticking—not just for Zak, but for all of us.

PART I

THE BROTHERHOOD OF SANGIN

1

Arrival

ZAK

AS MY BROTHERS and I walked into the madrassah, the man we simply called "Teacher" seemed unable to contain his energy. His eyes, rimmed in the kohl black makeup the Taliban loved to wear, were wide. I was only eleven years old, but I had been living under the Taliban for five years and had learned to pay attention to their moods. Teacher was usually very calm as we came in, sitting cross-legged as he waited for us to settle on mats in front of him before leading us through our lessons. Today he stood, shifting from foot to foot. Smiling more than usual and twisting his hands together, it seemed he had something exciting to tell us. I looked forward to anything different from reciting the Koran for hours, learning the *surahs* by recitation. I hoped we would play a game.

"Young brothers, sit down! Today is a great victory for Islam! Today is a great day for the Taliban over the infidels in the West! In the night, Talib martyrs in their jets struck the center of evil in America and toppled their buildings, by the grace of Allah!"

As he continued speaking, describing airplanes and fiery justice, I understood him to say Taliban jets had destroyed some important buildings in America. Teacher spoke of massive towers, marketplaces where Western infidels and apostate Muslims worshipped money over the word of Allah.

I did not really understand what Teacher was telling us. I came to understand the horrors inflicted upon America on September 11, 2001, by living the results in my own country and there coming to know and love American Marines as I love my own brothers. But that day, I was just a boy growing up in Asadabad City, of Kunar Province. I had never been beyond the city, where a three-story building was a tower. I had no real way to understand the meaning or importance of what our teacher was telling us. The Taliban made it that way. I did not know where New York or Shanksville or Washington, D.C., were, but I knew we were expected to celebrate attacks upon them.

I had lived under Taliban rule since I was five years old. I could not remember a time when they did not control Afghanistan. I was unsure how their victory in America would change anything. But I would learn soon enough that the grim expression on my father's face as he listened to the BBC on a prohibited radio the night before was connected.

My name is Zainullah Zaki. You may call me Zak. I am Afghan. I am Pashtun. There has never been a day in my life untouched by war.

IT WAS OCTOBER 7, 2001, almost a month later, when I awoke to the stars still overhead and the sounds of my two brothers, Asadullah and Faizi, snoring lightly. The sun was just an idea coming over the mountain, but I knew the call to prayer would soon come from the mosque. Our one-year-old brother, Zaidullah, slept with my mother and father in their room. I could hear my father, Ahmadullah, talking to him quietly, saying the nothings you say to a baby as he prepared to leave for his walk to the mosque for *fajar*, the morning prayer. My brothers and I prayed at home most days. We only made the trip to the mosque on Fridays.

My brothers and I slept in one of four rooms in our family's home, built of stone from the mountains towering over us. Our door opened directly onto a common terrace; all of the small parcel was enclosed by a stone wall. Leaving my bed, I stepped outside into the purple before the dawn. The air was still refreshingly cool but would soon begin to warm. Our cow and sheep were beginning to stir in their room against the back wall of our family compound.

I went to the bathroom to begin my ablutions in preparation for the first of the day's five prayers, passing the kitchen where my mother, Fatima, was already at work. She rose early every morning to complete her own prayers and milk the cow and sheep before preparing our morning meal of fresh bread, called naan, and hot green tea. Our kitchen was away from the sleeping rooms to prevent fire from spreading from the stove.

"Good morning, Mother." She looked up from rolling dough and smiled at me. Her eyes crinkled at the corner as she replied, "Good morning, *zarrgiya*." Pashto for "my precious little heart."

Asadullah soon followed me in and began his ablutions as well. He looked at me and laughed at my hair, still a mess from sleep.

"Don't laugh at me! You are late and Faizi is still asleep!" Asadullah was fifteen; Faizi was nine. We heard the bang of the front gate that meant my father had left to pray. Asadullah smiled.

"Let him sleep."

I liked being alone with my older brother. I was very proud of him. We walked out onto the terrace of our home, surrounded by strong rock and cement walls. The buzzing of the speaker at the mosque coming to life echoed down our streets. We are Pashtuns. We speak Pashto. But the language of Islam is Arabic and the call to prayer, called the *adhan*, is in Arabic. *Adhan* means "to listen." We faced west toward Mecca, unrolled our prayer rugs, and knelt. Faizi ran outside and unrolled his own rug beside us. We raised our hands, palms up, then knelt and prayed as the *muadhan* called to us four times.

"*Allahu Akbar.*" God is great.

Then twice each, "*Ashhadu an la ilaha illa Allah.*" I bear witness that there is no god except the One God.

Then, "*Ashadu anna Muhammadan Rasool Allah.*" I bear witness that Muhammad is the messenger of God.

"*Hayya 'ala-s-Salah.*" Hasten to the prayer.

"*Hayya 'ala-l-Falah.*" Hasten to success.

"*Assalatu khairum-minan-naum.*" Prayer is better than sleep.

"*Allahu Akbar.*" God is great.

"*La ilaha illa Allah.*" There is no god except the One God.

With morning prayers complete and my father returned from the mosque, we ate. Then there was just enough time to help my father before heading to school. He stood and beckoned to us.

"Come, boys, we go to the fields."

Asadullah, Faizi, and I followed my father to our small farm plot to help him with his labors. Like most people in our village, we were a poor family. Everyone worked except my baby brother. After an hour working, we ran to clean up, dress in the white clothes required by the Taliban, and put on a hat for the walk to school. We had to wear a hat at the madrassah or else our Taliban teacher would punish us.

Our teacher was still talking about the Taliban's great victory in America. For the last month, he had told us how the Arabs and their leader Osama bin Laden helped the Taliban attack the infidels, as they had done in the war against the Russians. He explained that because of what the Taliban and Arabs had done, the *Ummah,* the whole Muslim world, was winning the war between believers and infidels. Though he celebrated the triumph, we learned nothing more about the details of it. In school we still focused on reciting the *surahs* of the Koran. We memorized them word by word. The Taliban taught us little else. There was nothing else they considered worth knowing.

Line by line, he read and we repeated:

"Allah! There is no God but he, the Living, the Self-subsisting, Eternal. No slumber can see Him nor sleep. His are all things in the heavens and on earth. Who can intercede in his presence except as he permitteth? He knoweth what appeareth to His creatures as before or after or behind them. Nor shall they compass aught of his knowledge except as he willeth. His Throne doth extend over the heavens and the earth, and he feeleth no fatigue in guarding and preserving them for He is the Most High, the Supreme in glory."

We ate lunch in the madrassah, naan and tea again. Since other families were as poor as ours, an extra meal was enough reason for many boys to come there. I now knew there was a world beyond Afghanistan, but there was no real way for me to access it. Television, or any other kind of media expression, was prohibited in 1996 when the Taliban took control of the country. Radios were kept hidden, brought out from hiding and pressed to ears late in the evening, as my father seemed to do more and more. There were no broadcasts we could hear coming from Afghanistan anyway, unless they were Taliban controlled. My father listened to the BBC on a radio station broadcasting from Pakistan, the border just eight miles away. He had heard of the attack on the towers that way a month before. But nothing much had changed in Asadabad since then. The Taliban were still our teachers, our police, our masters. There were still no civil rights or courts; nothing resembling what I would ultimately come to know as justice. Ignorance was a virtue. Cruelty was common.

When we got out of school in the late afternoon, there was always more work to be done. But we were children and the mountains standing above called us. Houses were built into the hillside at their base, with one house's roof sometimes forming the terrace for another. With the Kunar River flowing through our village and the forest surrounding us, there were many things for boys to do.

Asadullah is the dutiful one in our family. He always thinks of his obligation to our family and others before himself. I was surprised when he said, "Let's go to the *jangal* before we go home." In English, you call a *jangal* a forest. I loved to play there. "Good idea!"

He smiled as he said, "I'll race you!" He took off running. Faizi and I were younger and slower. We could not keep up with him even in a fair race. This time he was already down the street and I could hear him laughing at us.

I yelled, "It's not fair!" but he was already too far gone. I looked at Faizi and smiled as Asadullah had. I was laughing as I said, "I'll race you!" and took off, too. Faizi was close behind me when we reached the banks of the Kunar River flowing near our house.

We all loved to swim across the river and play football, volleyball, and cricket in the forest on the island in the river. October is a beautiful time in Afghanistan. Though the water was cool enough to take my breath as we dove in and swam across, the weather was mild, about 65 degrees. Our wet clothes felt cold on our skin as we played late in the forest. It felt too good to be free to go home.

When I got home that evening, the sun was going down. It was a night like any other, but the atmosphere felt different as I came through the door. My father sat near the fire, his small transistor radio pressed to his ear. He looked shocked.

I asked Asadullah, "What has happened? Why does Father look so afraid?"

He shrugged his shoulders and looked at our father for information. Our father signaled us to be silent as he finished listening, then dropped the radio to speak to us.

"The Arab, Osama bin Laden, is on the radio. The Americans are attacking Afghanistan. The Taliban are running."

I wondered what Teacher knew of it. I wondered if I had to go back to the madrassah. I could not imagine life without the Taliban but I knew joy in that moment.

Down the hill from our house, I heard a man shouting. At first I did not know what he was saying. But soon I understood.

"*Allahu Akbar!*" God is great!

In my house, we knelt and prayed for peace.

ON THE DAY we now call 9/11, I was one of about 500,000 people living in Kunar, one of Afghanistan's thirty-four provinces. Kunar forms part of Afghanistan's eastern border with Pakistan. The province is very beautiful. The steep mountains of the lower portions of the Hindu Kush cover almost all of it. The few flat areas are for crops. Our agricultural fields were also terraced. They look like beautiful green stair steps when the crops are growing, mostly rice, sugarcane, wheat, and the vegetables we ate and also sold in the market. The mountains are cut by the Kunar River, which gives its name to our valley. The Pech River tumbles through its own valley to join the Kunar River at Asadabad City, the capital of Kunar Province and about eight miles from the border with Pakistan.

The villages in places throughout Kunar, like the Pech Valley, are hard to access. The people are Pashtuns like me but in the valleys the people like to be left alone. They are loyal to their villages and their tribes. They don't feel much connection to a central government. Still, there are common cultural connections. My father and mother taught them to me as Pashtun fathers and mothers throughout Afghanistan must.

"Zainullah, we Pashtuns live by *Pashtunwali*. The code requires us to live by specific principles. People most often know *Pashtunwali* for the hospitality required for visitors and the asylum offered to anyone who asks."

There is also a code of revenge. Many times, my father reminded us, "Pashtuns do not tolerate insults. We are expected to take revenge or negotiate compensation. We must show courage in our daily lives and bravery in defending our communities."

My mother always said, "Pashtun people must show loyalty, kindness, and respect."

Under *Pastunwali*, disputes are settled by arbitration at the *jirga*, a gathering of respected members of the community. Our faith, and our pride in being Pashtun, mean we take matters of female honor, personal honor, and defense of Pashtun land very seriously. Geography and *Pashtunwali* make Kunar's remote villages very good places to live if you are hiding. They have been good places for insurgents for hundreds of years.

Before I was born, Kunar Province was one of the centers of the resistance to the Russians when they were still the Soviet Union. The war against the Soviets, the foreign fighters who answered the call to jihad, and American support for them are what created an eventual base of support for the Taliban decades later. Being so close to the border with Pakistan made Kunar a good place for the mujahideen, the fighters who fought the Russians, to cross into Afghanistan. They used paths through the very high, steep mountains to attack and then go back across the border. Kunar is also where many of the foreigners who came to Afghanistan to fight the Russians hid. Some stayed after the Russian war, living far back in the valleys. The isolation of the villages in the valleys, and the extremist madrassahs on the Pakistani side of the border, made it easier for the Taliban to gain a foothold in those areas when they arrived almost three decades later.

The American ambassador to Afghanistan, Adolph Dubs, was kidnapped and murdered on February 14, 1979. On December 24 of that year, the Russians invaded to support the communist Khalq government, which had taken control of Afghanistan by a coup in 1978. Afghans fought for ten years and drove out the Russians in 1989. When they left, Mohammad Najibullah Ahmadzai was president. People called him Dr. Najib. The Russians supported his government even though he did not run it as a communist state and declared Islam as the state religion.

My family was part of the Najibullah government and that of Babrak Karmal before him. My mother's father worked in the office of President Najibullah. She was raised in Kabul. When she and my father got married, my father was a soldier fighting for the government in Kunar. Her father brought her from Kabul to Kunar to marry my father and that is where they stayed.

After the army, my father worked as a clerk for the International Federation of Red Cross and Red Crescent Societies during the time of Dr. Najib, trying to help with the effects of years of war upon the people of Afghanistan. *Pashtunwali* demanded we act with charity and kindness toward any people who needed help. Just as he taught us the rules of *Pashtunwali*, my father taught us that Afghanistan was a nation, not just a collection of tribes. And in a nation always at war, there was a lot of work to be done. Food. Housing. Treating the wounded. That was before he attended to the work of being a husband and father to a growing family. Still, he reminded us, "Never forget, you are Muslim, you owe *sadaqa*." *Sadaqa* is charity given voluntarily in order to please God. Sometimes even a smile is *sadaqa*.

Even after the Russians left, there was war. Dr. Najib's government collapsed in 1992. Then in 1996, the Taliban came from Kandahar Province, far away in the south. When the Taliban captured Dr. Najib in September 1996, along with the nation's capital city, Kabul, they tortured and killed him. They dragged his body through the streets of Kabul behind a truck. Then they hanged Dr. Najib and his brother from a light pole in front of the presidential palace so everyone would know the Taliban was in charge.

Mullah Mohammed Omar led the Taliban from a palace in Kandahar. The Talibs called Mullah Omar "Commander of the Faithful" and the country he ruled the "Islamic Emirate of Afghanistan." Under him, the Taliban massacred people who believed differently about Islam. They targeted Afghan ethnic minorities like the Hazara, Tajiks, and Uzbeks with killing, forced migration, and rape. They destroyed beautiful cultural landmarks that were part of our history, like the giant Buddhas in Bamiyan Province.

I was only five years old when they came from Kandahar, with their long beards and black kohl makeup surrounding their eyes, to take control of Kunar, but I knew that my parents were scared. I remember life was very hard when they were in power. In Kunar, we Pashtuns were already very conservative. But the Taliban declared controls over everything in our lives. Each night, I heard father tell my mother about new rules imposed by the Taliban. Every day there were new prohibitions.

"They have declared television and music un-Islamic!"

Then, "They closed the newspaper!"

Then, "Women may no longer work or go to school!"

Father began to grow his beard. My mother had to wear a burqa, covering her entire body, outside of our home. Even her eyes were hidden behind a screen. She was prohibited from looking at strangers and could not go out of the house without my father or one of her sons. It was strange to me that I had to escort my mother.

Except for the madrassah, there were no schools, no universities. Electricity was scarce. Maintenance of roads and buildings stopped. There was really no access to any of the basic life needs a government is supposed to provide.

Except guns.

There were always plenty of guns. The Talibs carried AK-47s and rocket launchers everywhere they went.

There was sharia law, of course. That was a form of government the Taliban were interested in. They cut off the hands of thieves. Talibs from the "Department for Enforcement of Right Islamic Way and Prevention of Evils," the religious police, enforced laws about modesty with sticks. They beat people in public who they said were not following the laws, especially women. Many people were killed by beheading or stoning as punishment for their violations. There was no real court, just the Taliban.

Although they were from a very different part of Afghanistan and had very different accents, the Taliban are Pashtun like the people in my village, so some of our neighbors were for the Taliban. My father was not and he suffered for that. Power shifts suddenly in Afghanistan. Who you know and support matters more than who you are. It has always been that way.

My father lost his job because he had worked under Dr. Najib. He grew his beard ever longer so he would not attract Taliban attention. Without his clerk job, he became a farmer. In our family, any child old enough to walk was a farmer. We grew enough vegetables to feed ourselves with a little left over to sell in the village market. We grew and sold flowers, too, red roses. Afghans love flowers and plant them everywhere. Maybe they represent hope that something good can grow from dust.

I WAS ONLY eleven years old on September 11, 2001. I did not really understand what happened that day. We really had no concept of America or the West. We were told that the people there were infidels and that they would try to make us stop practicing Islam or living in an Islamic way. I certainly did not understand their governments or laws or the things to come soon. After the Americans began their invasion, I still farmed and played cricket in the forest, but our village was growing more and more frightened every day. People were scared that the Americans would kill us with their planes, especially those who worked for the Taliban or Al Qaeda. My father had the radio up to his ear all the time.

The day American bombs started falling across Afghanistan, the president of the United States said, "The oppressed people of Afghanistan will know the generosity of America and our allies. As we strike military targets, we will also drop food, medicine, and supplies to the starving and suffering men and women and children of Afghanistan."

The world had finally remembered we existed. The Americans came with more guns and more trucks and more planes. We hoped they would bring good things, too. Afghanistan had nothing. After two decades of war there was no government. We had no schools. No police. No military. Literacy was 18.6 percent in 1979. In 2001, no one knew because no one had asked. There was nothing to make us a nation except our hopes.

Things seemed to be changing so fast at that time. Suddenly, the Taliban were gone. They moved deep into the more remote valleys like the Pech and Korengal, to places like Nangalam, Waygal, and Watapur. When I saw my first Americans, I was in Mandakol, a village near Asadabad. The Americans landed in what I would later recognize as an MH-47 Chinook helicopter. They were walking through Mandakol on the road. I was playing football with my friends. As they approached, we stopped and stood and watched them come. I remember being so scared of them. I was frozen, too scared to run. We all were. I just stood still. I felt like I could not move. They were watchful but they did not threaten us. They waved to us. Now I think they were as curious about us as we were about them. The soldier nearest to me was huge in his helmet and body armor, with sunglasses and guns. I had never seen such things before. I was just a boy in a traditional *shalwar khameez*. He smiled at me, but it seemed like someone had arrived from space. Soon there were more and more. I learned that they brought hope for a brighter future for us.

In the years after the invasion, our family was still poor, but the economy and productivity of Kunar were improving. The Americans helped the new government grow. Throughout the province, they built schools and medical clinics and civic buildings. They helped businesses succeed by bringing cell phone networks and developing roads and bridges to connect the district centers of the province. People opened new shops every day. My parents had five more children!

All of these projects created jobs for people, especially in construction. Of course, in Afghanistan, there is always corruption. People paid bribes for jobs. People paid bribes for licenses and project approval. People paid bribes for anything that someone else controlled. But I was a boy. I did not have to bribe anyone for the thing that mattered most to me. For the first time in my life, I was attending school. It was in the teacher's house, but finally we could learn. We still learned the Koran, but I will never forget the feeling of freedom as we learned other things we needed to live every day in the country we were building: English, biology, mathematics, and basic computer skills.

I began to see a future for myself beyond our fields.

2

A Hippie Cop's Son

TOM

AT 8:30 ON a perfect Tuesday morning in Chicago, Brother Brennan wheeled a television cart into our theology class at Marist High School. I was only fifteen years old, but I could read the stricken look on his face. Something big was happening. The energy he exuded told me it was something bad.

It's funny the things you remember: the grim set of his mouth; the way the cart's power cord was wound in a figure eight around the cleat bolted to the side of the cart; the way his shoulders seemed to slump for a moment, then reset as he gathered himself. Brother Brennan turned back to us, the screen coming alive behind him. It was an old television and it had to warm up before the screen came alive with a now-iconic image, one burned into my mind's eye: the top twenty floors of World Trade Center Towers One and Two pouring smoke from stark black holes ripped in the face of those soaring buildings, flames licking their edges, steel beams hanging from them like broken teeth from a bareknuckle fighter's mouth.

He seemed to stumble, looking for words he was ill-prepared to deliver. When he spoke, his words came haltingly.

"Gentlemen . . . I have to tell you something. Our nation is under attack. From whom I do not know. But our faith in God will carry us through this, as in all things."

We watched smoke pour into a cobalt blue sky. I was frightened by the image on the screen. Questions emerged from the confused silence. One of my classmates asked Brother Brennan, "Are we at war?" Another asked of no one in particular, "Who would do this?" There were not a lot of dinner table conversations in my house. When there were, foreign policy was not a topic. But I don't think anyone at Marist knew any more than I did. I had never considered that anyone might not like America. I had never really considered anything outside of America, period. I wanted to talk to someone, but I didn't know what to say. Looking around at a class full of equally stunned tenth-grade boys didn't offer many options for solace or wisdom. Brother Brennan seemed no less frightened and bewildered than the rest of us. Looking back, I am sure he had no concept of Al Qaeda or the Taliban but he tried to put on a show of composure for us. He was partly successful. Like all of us, he just sat with his sadness in the midst of our collective impotence. "Men, I think we should pray." He led us in the Our Father.

"Our Father, who art in heaven, hallowed be thy name. Thy kingdom come; thy will be done, on Earth, as it is in heaven. Give us this day our daily bread, and forgive us our trespasses as we forgive those who trespass against us; and lead us not into temptation, but deliver us from evil."

When he was done, I kept my eyes closed and prayed the Hail Mary.

"Hail, Mary, full of grace, the Lord is with thee. Blessed art thou amongst women and blessed is the fruit of thy womb, Jesus. Holy Mary, mother of God, pray for us sinners, now and at the hour of our death. Amen."

Although I'm not Catholic, in the second of my nine years of Catholic education, I didn't have a better idea.

On the television screen, people leapt from a tower's wounded visage before the camera cut away. I was horrified and confused, but I wanted to help; to respond in some way. I've always felt called to serve. Marist High School's mission was to reinforce that drive to be there when it matters.

As I sat in my tenth-grade theology classroom, I made a quiet vow to myself that I would serve our nation. I felt a righteous anger calling me to serve our nation somehow, but I certainly didn't know then that tremors from the collapse of the Twin Towers would reverberate as personal cataclysms for decades to come.

The Marist administration must have felt as rudderless as I did. They released us early and told us to go home. Mine was too far away so I walked to my aunt Jody's house and used the hidden key to let myself into an empty house. In the oppressive quiet, I felt a deeper fear and loneliness than I had in theology class. Then I saw her Bible.

My faith has been the single most important thing in my life since I was fourteen years old and fell on my knees in prayer to arise born again in God's grace. There's my life before confessing my sins, accepting forgiveness, and believing in Jesus Christ the Savior, and everything after. As I picked up the Bible and opened it to Psalms, I prayed for direction and reassurance. When I opened my eyes, I was still scared, but I knew that God would give me a path to help, even if that path was unclear in the moment.

PRIVATE SCHOOLS TYPICALLY attract kids whose parents have the money to pay for them—lawyers and doctors and investment managers. Mine was anything but that. My big, loud Greek family produced a hippie mom turned Chicago cop who raised my sister and me on the south side of Chicago in a blue-collar Irish-Catholic neighborhood full of cops and firefighters. Meanwhile, my father lived eight hundred miles away in Georgia with a felony record.

Our house had an ever-changing cast of relatives passing through as long-term guests. I grew up steeped in chaos, surrounded by weak men who wouldn't control their temper, didn't consider consequences, and bullied women through their body language or by getting even louder than loud women. My mom did what she could to provide stability, but in my childhood memories there is a through line of angry screaming, holes in punched walls, and thrown glasses shattered. The effect of that has been for me to seek the opposite in all things. It is almost incalculable to me that someone does not consider consequences. I am very comfortable in chaotic situations, but control over myself and my surroundings is second only to my faith in Christ in defining the essence of who I am.

My mom was a good kid left to fend for herself. The south side offers a million ways to get off track and by thirteen she was feral, drinking and drugging for the remainder of her teenage years. At seventeen, she was waiting tables in Chicago to pay for partying when my father showed up on a motorcycle with a guitar and long blond hair. He chatted her up and played her a song. She put down her order pad, hopped on the back of the bike, and rode off with him. Then he decided they needed to move to Georgia.

By eighteen Mom was a pregnant high school dropout who quit using out of duty to me, her unborn child. Three years later, she was married with two kids, living eight hundred miles away from her family. Things got bad fast. My earliest memories are of them fighting. When my mom confided in my father's sister, "I have to get out of here, I can't raise these kids in all this," she agreed to take the three of us to Chicago. Mom packed three-year-old me and my three-month-old sister, Jessie, into my aunt's yellow Datsun hatchback and ran back to the south side. When we got there, we landed at my great-aunt's house. Mom called my father in Georgia and told him we wouldn't be back.

The first year in Chicago, my mom, my sister, and I lived in one bedroom at my great-aunt's house: Jessie in a crib, Mom and me in the bed. My grandma and my great-aunt shared a room; my great-aunt's kids Meghan and Paul, were in another. Assorted cousins passed through as required, with an uncle or cousin usually crashed on the couch.

Over that first year back in Chicago, my mom cleaned enough houses and waited enough tables to get us into a roach-filled apartment building across the alley from a Chinese restaurant. I still can't escape the smell of the garbage cans. By the time I was six years old, waitressing was bringing in enough money that she could stop scrubbing other people's toilets and just bring them their drinks. She was working till two or three in the morning, so Jessie and I usually slept at my aunt's place again. In the morning, Mom would pick us up and get us to school, then go home to sleep, pick us up and feed us, and start the cycle again. I think about her when I tuck my own kids in at night and thank God I don't have to leave that night.

In a way, the bars Mom worked in gave us everything that came after. The table and chairs in our apartment were hand-me-downs from one of the bars where she worked. Mom bought our couch with five hundred dollars she won in a "Hot Legs" contest at another. Then, after another late night had become another early morning, a girlfriend talked to her as she counted out her tips.

"I'm taking the Chicago police exam tomorrow, Grace. It's a city job with benefits. You should come." Mom laughed.

"Hahaha! Are they looking for high school dropouts who hate guns?"

"You only need a GED, Grace! Let's do it!" Mom had had the foresight to get her high school graduate equivalency degree in Georgia. The next day they both headed to the high school from which Mom had never graduated. Ten weeks later she got a letter from the Chicago Police Department telling her she was "well qualified" and that, if she wanted the job, she needed to buy a gun, belt, nightstick, and armored vest because the city didn't provide them. My mom, the hippie chick who started drinking and using drugs at thirteen, the single-mom child of a single mom, became a Chicago police officer for the health insurance. She had to take out a loan to buy the equipment. From her I learned that you have to be relentless in pursuing things that matter, especially when they make the world a better place for the people you love.

THE CHICAGO POLICE Department was a challenging place to be a woman in 1992. Not everyone wanted a woman on the force, much less in their patrol car. She was a petite woman with long blonde hair who had to endure a lot. At five foot two and 120 pounds, Mom broke out in laughter the first time she looked at herself in the mirror in uniform. Despite her size and the fact that she was only twenty-four, Mom was already tough as a result of more than a decade of fending for herself. I was proud of her. I still am.

My great-aunt, whose house we lived in when we first came to Chicago, and my grandma, who had not always been on the right side of the law, were totally on board. They watched Jessie and me while Mom was in the academy and then whenever duty demanded it. They made sure we got to school and anywhere else we needed to be.

Police work got us into another, better apartment, away from the garbage cans. Day shifts came with the badge. Mom worked while we were at school and was home at night. But most days, after patrolling some of the most dangerous streets in America, Mom still cleaned our eighth-floor apartment building's halls and porches to get a break on rent. A lot of the time, it felt like we were both getting raised by the TV, but I was also cognizant that what she was doing was for us. Mom would come in and change out of her patrol uniform, then head to the kitchen to make us dinner before going downstairs to push a mop. Whether she was putting herself at risk dealing with murders, drug abuse, and domestic violence cases to make sure my sister and me could see a doctor when we needed to, or mopping hallways, she cleaned up other people's messes to keep a roof over our head. My mom's example taught me about character and toughness, resilience and commitment. Most importantly, she taught me about service.

Despite the chaos of our family, I always knew I was surrounded by people who loved me, particularly my mom. Every night, I climbed into Mom's bed and we read together. Mom almost always chose stories about epic journeys against overwhelming odds. It's not hard to see through that one. Propped against pillows, we worked our way through *The Chronicles of Narnia, The Hobbit,* and the *Lord of the Rings* trilogy. I realize now how exhausted she must have been, how she had to fight to stay awake while she read to me and how much she loved me and my sister to endure the things she did. As the day came to a close, when she had finished whatever she needed to do to make sure we had food and shelter, she affirmed to me, "Thomas, never forget, you're a beautiful boy who can do anything."

We didn't just read. To an extent, my mom talked to me like I was an adult. I could not understand the realities of her days working a dangerous job on a bad side of the city. But she needed to unload with someone not on the job. Most nights, we talked when she got home. Her day, my day. When I think about it now, I see her hands, chapped and red from the cold winds out on the street, further inflamed by the hot water from washing dishes and getting Jessie bathed. She would tell me what she could about her day in a job that meant drug overdoses and domestic violence calls. Once it meant she got cut up pretty badly in a brawl with a guy that left her with lasting health implications. But what could she really say?

"Thomas, today I wrestled a guy almost twice my size who was beating his wife with a curtain rod when we got to his apartment."

"Thomas, a heroin addict puked all over me today."

"Thomas, by the time we arrived at the shots-fired call, there was nothing left to do but watch the guy die and try to keep his kids from seeing him do it."

Mom was on one side of the law; my father was on another. When I was eight years old, my father went to prison in Georgia. He called me from prison not long after that. "Hey, buddy," he said, "I have to take a trip for a while. I won't be able to see you."

I hung up on him. It wasn't that I didn't want to talk to him. He certainly wanted to talk to me, to try to explain the situation to me. But I was just overwhelmed and I didn't understand why I could not see him. I loved him and wanted a dad in my life so badly, but it was too painful to almost have what I wanted without being able to really see him. I did not talk to my father again until I was thirteen years old. But it's worth noting that he called and wrote me letters for five years. He kept trying and he's a part of my life now because he did.

The fact is, I needed a dad in my life then. I listened to other kids at school talking about the things they did with their dads and felt a palpable emptiness in my center that no video game or pair of shoes could fill. I had nothing to contribute when those stories came up.

It finally came to a head one night. Mom was so tired she was moving at half speed as she made dinner for the three of us. Her hands and face were red, blasted raw from the winter wind on the streets. She still had to get us bathed, tuck us in, and clean up the mess two kids make before she could go downstairs to mop the building lobby. I tried to be a good kid. I knew she was giving us everything she had. But I was ten years old and that night all of her everything didn't matter. I was a child; overwhelmed and sad and alienated from my friends, even the kids whose parents were also divorced. You want to know if you can feel an absence? Yes, you really can, and it hurts.

I followed her into her room after dinner and sat there while she tried to decompress from her day. She was folding clothes as she tried to talk to me. I was nonresponsive. She looked straight at me.

"Thomas, tell me what's wrong." I was suddenly crying so hard I couldn't stand.

I lay on her floor. "I just want a dad," I said over and over through gasps and sobs.

Eventually, my best friend's dad started taking me to father-and-son events. He wasn't my father, but at least I was with a man who cared about me. I won an autographed picture of Chicago Bears wide receiver Tom Waddle at one father-son banquet. Winning that eight-by-ten black-and-white head shot capped off the best night of my life. Then my best friend's dad hung himself in his garage.

My father was in prison three states away, my supplemental father figure killed himself, and most of the men in my life were unaccountable. It made me mistrustful of men, and it also left me such that now when I perceive a void or gap, I feel a compulsion to fill it, to make it better. Put less positively, I am unable to resist trying to bend a situation to my will. I've never been the smartest or most talented guy in any room I've been in, but I am relentless—for better, or for worse.

MY FATHER GOT out of prison when I was thirteen. I started going to see him in Calhoun, Georgia, during my high school summers. My mom felt betrayed by my decision, but I needed to know my father. Culturally, Calhoun is about as far from the south side of Chicago as you can get. Suddenly I was in a city in the Appalachian foothills of northwest Georgia that had fewer than ten thousand people living there. I was living with a father I hadn't seen in almost six years. He was remarried and had another child, my sister Erin. It was jarring. But it was also everything I wanted. My dad and I rode horses and four-wheelers. My cousins and I rode jet skis at a nearby lake. It wasn't all fun and games, though. Once I turned fourteen, my father expected me to work in his auto body shop or in his barn.

"Thomas, there's a pile of gravel needs spreading out in the barn." That meant a ton of gravel that had been dumped had to be spread, shovel load by shovel load, to cover the barn floor. It was a summer's worth of twelve-hour days.

"Tomorrow morning, I have some vegetables for you to plant." That meant hand tilling an acre of rock-strewn Georgia red clay in 98 degrees and 94 percent humidity, then planting and tending a garden.

When I was sixteen, he told me I had saved enough money for a brown 1985 Nissan Maxima station wagon with 225,000 miles on it. My father put an aftermarket CD player and speakers in it. It was awesome, something for which I had worked and my father had provided. I started thinking about work then, about the dignity inherent in it.

Years later, as a student at the Marine Corps' Basic School for new lieutenants, we were in the defense, digging in to hold terrain against other student lieutenants, who were on offense. I was a machine gunner assigned to a spot right next to a gigantic tree I knew would have an insane root network. The expected standard was a chest-deep fighting hole. All I had was a pickaxe. I thought, "I don't know what I'm going to do." When I don't know what to do, I just start working.

My instructor walked by the first day and asked, "What are you doing, Schueman?"

"I'm digging a fighting hole, sir."

"You think you can get a chest-deep fighting hole in this kind of ground?"

I looked up at him. "I'm going to have a chest-deep fighting hole, sir."
He chuckled and walked away.

The next day I was at two feet deep when he came by. "How's that
chest-deep fighting hole going?"

"It's going, sir. I'm going to have a chest-deep fighting hole." By day
three, I got a chest-deep fighting hole. I was focused on one thing: digging a hole. I am not afraid to work hard. Some of that was definitely
honed hauling gravel in the Georgia summer sun.

BY SIXTEEN, MY life at Marist High School was a whirlwind. I played
football. But at the same six foot two and 140 pounds I had been since
seventh grade, when puberty showed up for everyone but me, I went
from starting every game to riding the bench. I had no talent and no
size, so again, I just tried to succeed through working hard. I practiced
like I was going to start every game. Between that and pickup sports, I
racked up an impressive string of injuries. I broke my nose three times,
my collarbone, my arm, and lost some teeth. I have a truly obnoxious
need to be the best at whatever I am doing. Football was not ever going
to be that so, in my junior year, I walked away from a sport I had been
playing since I was five years old. My coaches had given me a lifetime
gift of grit, but with college coming and no way to pay for it, I needed
to focus on academics in hopes of scholarship money.

I was even competitive as a nerd. I was in seven clubs. I was a member
of the National Honor Society. I was a drama kid. I was the captain of
the Speech and Debate Team, competing at state and national levels. I
spent the summer of 2001 studying constitutional law at Yale. A year
later I was at Stanford for more of the same. I was treasurer of the Service Club and president of the Ecology Club. I've always been attracted
to service, so I joined the Marist High School Mission Committee because I liked the service component and my friends were in it. I joined
the Ecology Club because I was in Advance Placement Biology, where I
shouldn't have been because I sucked. I actually do care about environmental issues, so it aligned with my values, but also, my AP Bio teacher

ran the Ecology Club and I figured it would help my grade. I joined the National Honor Society because that's what you do to go to college. Anything for the win. At the time, I couldn't have told you why I did all those things, beyond the notion of "college." There was no competition for best self-reflection. I was just desperate to get out of a years-long cycle of closed-off options and I assumed college was how people living the life I wanted did that.

I don't think my mom has ever pressured me to do any of the things I have done. She certainly never said I had to go to college. No one in my family had gone. I don't think anyone in my family knew *how* to go, least of all me. I was seventeen, sitting in the basement of our house, staring at college applications to places that looked like a fantasy a long way from the south side of Chicago. There's a tyranny in ignorance.

How many kids get left behind every year because they don't know how to move forward? Even if she had wanted to, I don't think Mom could have educated me about, or tried to prepare me for, the upward mobility I was intent upon. I knew I had to get out, but no one we knew had done it. I assumed that people went to college and got handed a profession. I did not understand anything beyond that and I had no one to explain the process, even if I had known what to ask.

I took the ACT and scored a 36 in English, which is a perfect score; I had a 17 in Math, which is abysmal. I still got into Loyola University–Chicago but I was clueless about how to make it happen. I was well past any scholarship deadlines when I started applying because, again, I had no idea how those things worked. I was insanely driven, but the most powerful boat needs a rudder. I had nothing except my willingness to keep pushing.

This powerful mixture of extreme competitive drive and a general lack of direction is what brought me to the military. I did not grow up intending to serve in the military. Like college, no one in my family had served before, but even more of an issue was that my mom actively forbid it. I was called to the military from a general desire to be of service to my nation. The summer before my freshman year, I learned that the Naval Reserve Officers Training Corps (NROTC) existed. Had I gone to the Army or Air Force ROTC office first, I would likely be there now. But I went to NROTC first.

The recruiter was straight with me: "Mr. Schueman, you are way too late to apply for that scholarship, but if you join NROTC and participate, you can compete for a scholarship next year."

My mom had already taken a second mortgage to get me through my upcoming freshman year, so a possible scholarship for sophomore year was like a life preserver thrown near, but not to, a tired swimmer. There was still some distance between me and salvation, but it made survival seem more likely. I did not have any role models to drive me toward the military. My mom was a daily example in her life as a police officer, but I knew that was not the direction I wanted to go. NROTC became an easy decision.

"Where do I sign?"

I still really wasn't thinking about what joining NROTC meant. I wanted a professional career in a white-collar environment. I wanted to somehow serve my community and country. Like everything else, I was completely ignorant of what all that really meant or how to get there. Beyond being a lawyer or a doctor, I probably could not have told you what other professions existed. I wanted to be a lawyer because my grandma was obsessed with Tom Cruise and I had seen *A Few Good Men*. I knew his character was a Navy lawyer. I knew lawyer was a career where people had opportunities for stable, upper-middle-class lives. I figured if I could get NROTC to pay for the rest of college, I would do a few years as a Navy lawyer and move on to environmental policy work.

I never rebelled as a kid. I never smoked. Never did drugs. But my mom only had two rules: no motorcycles and no military service, and I ultimately broke both of them. When I told Mom I was joining ROTC, she was furious. It was 2004. Afghanistan had been going for three years. Iraq was starting to come apart at the seams and she had not said the names Bush, Cheney, or Rumsfeld without spitting since 2003. I still was not paying attention to things like that. I just knew there was money to be had and we needed it.

"Thomas Schueman, I did not fly to Seattle to march or march here in Chicago to protest that stupid, illegal war so you could go get killed in it!"

"Mom, I will most likely be going into the Navy as a lawyer. I am not going to get killed."

Still not happy.

"You know we need the money!" I continued. "Jessie is only three years behind me and they said I will probably get a scholarship next year."

Still not happy, but now she couldn't argue. With two mortgages on the house and her holding two jobs already, there were no more options for college money.

In the end, there was just too much reality staring her in the face. She had worked so hard to get me to a position in which I could follow this intense drive I have. I had worked so hard to get where I was. To come up short of the mark now was simply inconceivable, even if it meant giving her son over to an institution which she not only had no faith in, but actively distrusted.

Off I went to the Naval Reserve Officers Training Corps orientation training at Great Lakes Naval Station. It was twenty minutes, ten miles, and an impossible distance from where she wanted me.

3

Clear Skies

BY THE SPRING of 2002, with the Taliban hiding in the mountains and valleys, more and more Americans began to arrive in Asadabad and throughout Kunar. To us, they were like people from another planet. They drove very fast, roaring by in big trucks and making a lot of dust on our narrow roads. At first it was the ones with the beards and sunglasses, the special operations soldiers. But soon American convoys were coming through our village all the time. They were trying to expand the government's influence north to remote villages hidden in valleys like the Pech, the Korengal, and the Waygal. The soldiers in the vehicle gun turrets threw us chocolate and toys. They had so many guns and so much equipment and food, it seemed impossible anyone could defeat them.

The soldiers raised tents. They filled sandbags and the big boxes they called HESCO barriers with dirt to make walls at an old Russian fort called Topchi. They surrounded the fort with coils of razor wire and called it Forward Operating Base (FOB) Asadabad. FOB Asadabad grew very fast. Helicopters and trucks came constantly, carrying material and equipment and, most of all, more soldiers. American soldiers began to build and move into new plywood barracks and headquarters buildings. They built observation posts (OPs) on top of the mountains surrounding the FOB and recruited Afghans to fight with them against the Taliban and Al Qaeda. Later they changed the name to FOB Wright. They named it after an American soldier, Sergeant Jeremy Wright, who was killed near there in 2005. Whatever the Americans called it, we Afghans still called it Topchi.

Slowly the Americans became familiar to us, and more and more Afghans started working for them. The Americans allowed Afghans to set up a market at part of the base. Some people sold them bootleg DVD movies, while others sold bread or souvenirs of Afghanistan. The American soldiers loved antique rifles, but most of the real ones were gone as souvenirs by 2004. Just over the border in Pakistan, people started making copies and the American soldiers bought those, too. After years of hunger and poverty, the Americans seemed to have everything anyone could want.

Their aircraft filled the sky. So did our kites. Kite flying is a team sport in Afghanistan. Wires controlling the paper kites are coated in a viscous glue impregnated with ground glass. The *charka gir* holds the spool or stick around which the kite's string is wound. The *gudiparan baz*, who flies the kite, seeks to use his own wire to cut that of the opponent team, sending their kite pinwheeling, adrift at the mercy of Kunar's winds coming off the mountains.

After five years' absence, kites danced across Asadabad's sky, a celebration one could enjoy on any day there was wind. On a day in which school had not yet resumed in earnest and many children were out for a *jang*, or fight, I watched my own kite flutter against an azure sky, a wire running through my gloved hand. My friend Latif held the spool of wire by which I controlled it. After years of uncertainty, the future seemed secure enough to simply enjoy the moment.

"Look out!" Latif saw a kite dancing close to ours. I twisted the wire, making our kite dart away from the opponent, then pulled again to try to cut the attacker's line. Neither of us could make sufficient contact to cut the other before a third kite dipped toward ours, its line suddenly tight. I could feel the impact in the taut wire playing through my fingers. Then . . . nothing. The wire spiraled downward in loops as our kite flew away, unfettered.

"Today is not our day!" I shouted. I heard Latif exhale through pursed lips. But when I turned to look at him, he was smiling. The Talibs were gone, the sky was blue, and the kites were flying. *Azadi rawest*. Free and legal.

The Americans were so rich, even we kids found ways to make money with them. The mountains were so steep and high that anyone on top could look straight down into FOB Wright. Afghan and American soldiers lived in the OPs on peaks surrounding it. They climbed up and stayed in the OPs for a couple of weeks at a time to keep the Taliban from launching rockets and mortars down into the camp. When they changed out soldiers manning the OPs, they had to carry two weeks' worth of food and weapons and equipment up with them. We knew their schedule, so we went to them asking for work. We did not speak great English and they spoke no Pashto, but we built friendships with the soldiers over time. We did a lot of signing and simple sentences.

"You give us job!" We rubbed our fingers together to simulate money.

They pretended to carry heavy bags. "You carry our bags up the mountain for us, we give you money!"

"Okay!" Both sides gave the thumbs-up, which we kids thought was very funny because it means the same thing to an Afghan as a raised middle finger does to an American.

Then we bumped fists.

The Americans always bumped fists with us or shook our hands or gave high-fives and called us "Brother." They were louder and looser than Afghans. We were kids and we thought they were fun. Most of them were only five or ten years older than us, almost kids, too. The soldiers going up the mountain paid us five or ten dollars for each bag we carried up from FOB Wright to the OPs. The ones going down the mountain paid us to carry bags down. It was a lot of money for us and helped our families very much.

To get there for work, Asadullah woke me even earlier than normal.

"Zainullah, get up! We must say our prayers and get to FOB Wright in time to get back down the mountain for school!"

The soldiers were weighted down by their armor and equipment as we climbed. They moved slowly and had trouble breathing. We teased them.

"You too slow!"

They had dirty responses.

"Fuck you, little man!"

We had been running up and down the mountains for our whole lives. It was like nothing to us. The money was like nothing to them.

WITH THE HELP of the U.S. government, people in Afghanistan started to breathe again. Only Pakistan, Saudi Arabia, and the United Arab Emirates had recognized the Taliban's Islamic Emirate of Afghanistan after they took control in 1996. The United States and other countries recognized the government of the Islamic Republic of Afghanistan on December 22, 2001. Now the most powerful countries in the world were helping found a new Afghan government in Kabul to fight the Taliban. A man named Hamid Karzai became the president of Afghanistan. The Taliban ran far back in the valleys and hills where the people who still supported them lived. I call them the Highlanders. It was like a new life started for us. I didn't feel the Taliban on top of me all the time. I felt lighter. We were becoming a real country.

There were new construction projects and developmental programs coming to Kunar, especially around Asadabad. People started opening shops. They built new schools for girls and boys. In Asadabad, many of the shops had roll-down doors that the owners painted red, black, and green, the colors of the country's new flag. Finally, people could get an education and find work.

Even though he had worked for Dr. Najib's government before, my father kept farming. He told me, "I am too old to find work in the new government. This new country is yours, Zainullah." My brothers and I helped him, of course, but we were working hard to have other options.

More and more of my neighbors went to work supporting the new central government. Our neighborhood in Asadabad felt connected to

Kabul. To an American that probably makes sense. The cities are only 115 miles apart. But it has never been the case for much of Afghanistan to feel connected by a central government. Many Afghan people associate their identity with their tribe or their ethnicity or the valley where they live. The Taliban were never effective as a central government. They had just been different groups of people with similar ideas but acting differently depending on who was in charge in different places. Now people were at least able to participate in their government. If you had a complaint, you could take it to the government. There were still problems like corruption and bribes and favoritism, but no one was getting stoned or beheaded or beaten in the streets.

For me, the best part of all of it was the chance to go to school.

President Hamid Karzai said in a 2003 State of the Nation speech, "Education is a high priority." It had not been any kind of priority for the Taliban, so anything was better, but having the government say we needed education was a good thing. The year before Karzai said that, our village teacher had come back from Pakistan. He had been there since fleeing the Taliban in the late 1990s to avoid being killed for apostasy. We didn't have an elementary school building yet, so my brothers and I went to his house to learn. Even if it was not an actual school building, from eight till noon almost every day we were in school. People who have always had schools to attend do not understand what a luxury it is to learn. After the Taliban and the madrassah, I felt like our new teacher was teaching us everything in the world.

"Boys, there is a big world outside of Afghanistan. You must learn how to communicate and work here and elsewhere. You must learn about the world and how it sees us if you want to get anywhere in life. You have the chance to have more than was available to your fathers if you work hard and learn."

I was very happy to walk with my brothers for the two kilometers along the canal to our teacher's house each morning. Of course, there were no English or computer classes when the Taliban was under control. But the Taliban didn't teach us basic things we needed to know about ourselves or the world outside Kunar, either. Now our teacher gave us classes in human biology and chemistry. We learned mathematics. And girls could go to school.

In Kabul, there had been secret schools for girls during the Taliban time. In Kunar, they did not even go to the mosque then. People risked death to teach girls. Now they went to a different one than us boys, but they could go to a real school for the first time since 1996. Today, as a father of three girls myself, I will never accept less for them.

I was thirteen when I completed the classes I needed to start Omar Khan High School. I was finally in a real school building. It was wonderful. Things that are expected in the West were luxuries for us. I had new green uniforms to wear to school. American soldiers and other groups visited us and helped us build desks and chairs. They brought donated school supplies from around the world, blue notebooks from the United Nations High Commissioner for Refugees, pens from people in Europe and the United States. I continued my English classes, but with more resources, I learned faster. I worked very hard to learn English. I believed it was the key to success in the new Afghanistan, the most critical skill in the new Afghanistan we were building. We felt support from all over the world.

The happiest day in my life was my high school graduation in 2008. I was eighteen and the second in my family to graduate, behind my older brother Asadullah. I was very proud. In Afghanistan, a graduation means a big party. Everyone was proud of me because it is a sign of achievement and improvement for the family, village, and country.

The party was the biggest event in my life. The men wore blinding white *shalwar khameez,* with brown or black or gray worn vests over the long shirt reaching to their knees. The women wore colorful dresses. Some were decorated with bits of mirrored glass to catch the light. The women in my family cooked huge bowls of rice and meat, platters of bread and vegetables. We men danced the *Attan,* a group dance formed of circles of men, moving faster and faster to the sound of drums and clapping. Under the communists, men and women danced together. That was not the case in Kunar, even with the Taliban gone.

I loved learning. After so many years of being denied so much information, after high school I wanted to take the national university entrance examination called the *Kankor* exam and go to Sayed Jamaluddin University in Asadabad. The Ministry of Education controlled the 160-question exam. *Kankor* determined whether you could go to a public university paid for by the government or would need to attend a private university paid for by your family. I spoke to my father about it.

"Zainullah, even if you pass the *Kankor*, we have your brother Habibullah and your sisters Serena, Sangina, and Parkha now."

My parents had grown our family to nine children since the Americans had come.

"There is no money for more school for you now. You must work to support the family."

Even with the Taliban in hiding, there was not really money for anything except survival. In Afghanistan you support your family before you support yourself. If I am making money and my brother needs it, I give it to him. My older brother was already helping to support the family. I had to go to work.

It was not hard to find work in Afghanistan in 2008. It was hard to find *safe* work that paid well. I wanted to make money, to help my family and be ready when it came my time to marry. But the Taliban was getting stronger, especially since the Americans had become involved in Iraq, and jobs that paid well were often in Taliban areas. That was true when I got a job supervising laborers for a construction firm. The firm that hired me was building a district center in Nangalam to hold a *jirga*, the traditional meeting by which Afghans make communal decisions. Nangalam is about twenty miles west of Asadabad, where the Pech and Waygal Rivers meet. It is the largest town in the Korengal Valley. The job there existed because Americans were trying to help the new government connect different valleys and tribes and languages to Asadabad and then Kabul to make them part of the government. American and Afghan soldiers were fighting, and dying, just to bring a better life to the people there. They worked with the new government to improve the road connecting the villages in Kunar with Asadabad. The government wanted to bring civil administration and schools to them and make it easier for them to sell their goods in the larger cities, so they tried to fund civic projects.

Developing governance was dangerous work for everyone involved because in Kunar, the farther into the valleys and mountains you go, the less connected to the government the people are. The people in places like Nangalam and throughout the Korengal Valley did not support the new government like we did in Asadabad. The Korengalis speak Dari and Pashto, the two primary languages in Afghanistan, but also sometimes speak their own languages, like Nangalami and Korengali. Some of the areas are so isolated that dialects differ by village.

I was in Nangalam in the Korengal Valley for seven months, living at an American base called FOB Blessing and constructing a government building. The security situation was very bad in Nangalam and places like it throughout the valleys in Kunar. Just before I went to the Korengal Valley, about two hundred Taliban attacked and overcame a combat outpost (COP) in the Waygal Valley just north of the Korengal. They killed nine Americans at a village called Wanat.

FOB Blessing was surrounded by steep mountains and pine forests. When I got there, my supervisor showed me around. He pointed as we looked out at the mountains surrounding us.

"The Taliban hide in the mountains and forest and shoot at us all the time. Mostly, they are not effective. But sometimes they use the DshKa, a Russian 12.7-millimeter machine gun that fires bullets the size of a man's finger. A DshKa bullet can split a man in half if it hits him. We put up screens and worked behind them to keep the Taliban from being able to accurately shoot at us. If they see anyone outside the screens, we take fire from rockets, mortars, and rifles."

Some of the villagers in Nangalam got shot. The same thing happened to some of our laborers. There was little hope for them when they got shot. There just wasn't enough medical care available and almost no way to get them to a hospital. If the Americans had a helicopter, the person might live. If they did not, or could not spare it, the person might die.

We were trying to help the people in the Korengal Valley have better lives, but they did not seem to care. American and Afghan soldiers talked to them in the daytime, trying to bring them to the government. They offered roads, schools, and jobs. But the Taliban were threatening them at night, saying if they cooperated, they and their families would be killed. It was hard to convince them of a brighter future when they were scared for their lives now. That was the way of the Taliban. They came in the night and made threats. They intimidated people to keep them from joining the government that wanted to give them a better life. They wanted to take our freedom to choose.

Even though I was making money, I did not enjoy my life in Nangalam. I had no regular way to contact my family. When I was able to use the work phone to call home, my mother begged me to return to Asadabad even though we desperately needed the money I was making.

"I want you home and safe, *Zargiyya*! I don't care about money, I would rather have my son!"

There was little food for us workers. I earned vacation time, but I could not take any time off because the twenty miles of road home to Asadabad was too dangerous to travel by myself or even in a truck with other workers. Taliban ambushes were at close range because the thick vegetation on the roadsides allowed them to creep down from the hillsides. They bombed the road, too. When soldiers used the road, they sometimes took hours to go just a few miles because of the explosives the Taliban hid on the road. I found that out myself.

I was driving a truck filled with rocks from a quarry site back to Nangalam when I struck an improvised explosive device (IED). The explosion buried me in sound and light, as if I were dead for a moment and already in a world of noise and flash. There was no pain, just confusion and a strange loneliness, as I was briefly transported somewhere far away from a road through a valley in Kunar Province.

The explosion blew the engine out of the truck and destroyed it. The weight of the rocks may have kept the truck from rolling into the river beside the road and helped save my life. I walked away from the explosion. Looking at the truck, a U.S. soldier asked me, "How are you alive?" I did not tell my mother and father; I just began planning to leave.

I wanted to go get out of the valleys and back to life in a city where the Taliban had less power, so after seven months in Nangalam, I quit and went back to Asadabad to work on my family's farm. The soldiers at FOB Blessing kept trying to convince the Korengalis to join the government. The Taliban kept intimidating them and attacking people who were just trying to make a better life. Eventually the Americans left the Korengal and Waygal Valleys. It was just not worth the cost to stay in the mountains as long as we could keep the Taliban out of the cities.

WHEN I RETURNED home, I helped grow and bring in our family's vegetables. I took what we did not need to the market to sell with the flowers. I tended to the animals. I kept studying English on my own, even though it was hard without any resources. But my brothers were learning English as well, so we practiced together, having conversations as we worked around the farm. I coached my younger brothers Zaidullah and Faizi. Asadullah led us all.

Zaidullah was just beginning his studies. I did not tease him for sentences like "The plant, they good for eat." I tried to explain the way Americans spoke: "The *rice* is good to eat, Zaidullah." Asadullah was studying English literature. He read Shakespeare and Dickens and told us the stories he was learning as we cut flowers for the market. I still had plenty of English to learn myself. Just as I had wanted to learn more than the words of the Koran, I wanted to do more than farm for survival. I knew English was my path to university. I wanted my brothers and sisters to be able to go as well. That meant I needed money.

I had studied hard in high school to learn English and create more work opportunities for myself. I worked to develop technical skills so I would be able to provide as a good husband and father when my family found a wife for me. I was working hard to survive and help my family do the same, but it did not feel like building the future I had envisioned since 2001.

Then I heard about a company at FOB Wright called Mission Essential Personnel (MEP). MEP was recruiting English speakers to work as translators/interpreters throughout Afghanistan. I was concerned that my English was not good enough to pass the test, but I walked to FOB Wright to try to take the test anyway.

At the FOB, I waited in a long line with many other people seeking work with MEP. We were all hopeful and anxious. The man next to me was particularly nervous, since by 2010, suicide bombers had become a fact of life in Afghanistan. "I hope they hurry up. This line is a target."

"I just came from Nangalam," I said. "We got shot at every day. This is nothing." I wasn't boasting. Asadabad was home and I felt safe.

"I don't care about Nangalam! A bomber attacked the district center in Nangarhar last week! The Taliban and Al Qaeda train them in Pakistan, even young girls! Then they send them to look for large gatherings where they can kill and injure the most people! Things are getting worse, not better like the Americans promised." I could not argue against that after my time in the Korengal. I didn't even tell him about my own experience with bombs. But I knew it would take more than the Americans. Without Afghans involved, there would be no Afghanistan.

"That's why I am here, man. We have to help make a better future for our families and our country, where people are free to live the way they want to."

After a little while, I started to get nervous myself, watching people passing on the street. Just waiting in line to apply for a job was dangerous enough. Everyone knew that working for the Americans paid the best because it was dangerous. Life in the new Afghanistan meant people had to be willing to take risks to have better lives for their families. If I would not die for it, how badly did I want it?

Finally, after a long wait, an Afghan-American came to the gate. He led us through screening, where they inspected me to make sure I was not a bomber myself. Then we went into a large room in a stone building. We sat down at long tables. The Afghan-American began to speak in English, then Pashto.

"Before you is a test made up of one hundred pages of English sentences. You must read them and answer questions about them. Do your best. You have two hours. If you pass, we will discuss the next step."

When I left, I did not feel I had been successful. Walking back to my house, I felt sad and did not expect to hear anything. Especially in mountainous Kunar, where farmland was hard to come by, everyone knew that a job with the government or the Americans was the best opportunity. I could not think about going back to Nangalam, but I did not want to farm. Still, every day, I woke up and did my ablutions and prayers, then went to work on our food crops. I sold flowers in the market. As days passed, I became more convinced that I had failed the test to be a translator.

After a very long week, MEP called to say I had passed! The man on the phone said that if I wanted a job, I needed to get to the American base at Jalalabad for a convoy to Kabul. Jalalabad is fifty miles southwest of Asadabad, in Nangarhar Province. Like the trips to and from the Korengal Valley, that trip was also dangerous. But my family needed money, so I asked a friend to take me to Jalalabad.

I had lived my entire life within sight of Kunar's mountains. I could count on them to always be there. The mountains did not care who controlled the government or what I believed. As we drove the Asadabad-to-Jalalabad Highway, the mountains reassured me that some good things will always endure. But we also never forgot that the Taliban used them to hide. I counted on the fact that the Taliban did not usually attack civilian cars. As we drove the Asadabad-to-Jalalabad Highway, we passed the mouth of the Babur Valley. My friend gestured out the window of his car.

"That is where the mujahideen always ambushed the Russians in the 1980s. Now the Taliban use it against the American and Afghan government forces. Are you sure you want to work for the Americans?"

"They are here helping us. I will accept what they will accept. I want to help my country improve."

That ended the conversation.

Once I got to Jalalabad Airfield, MEP put all the new interpreters/

translators in a convoy of vans and sent us to Camp Phoenix in Kabul, another ninety miles away. That was another dangerous trip. The Kabul–Jalalabad Road is also known as National Highway 08. It is part of the Grand Trunk Road built 2,500 years ago, connecting the countries we now call India, Bangladesh, Afghanistan, and Pakistan. In buses, the narrow road is frightening. There are steep drop-offs on the side. The highway runs parallel to the Kabul River and is often blocked for hours by car accidents and rock slides. The man sitting next to me was worried about other dangers. "Let us hope we do not find a Taliban checkpoint. We will be lucky if they only rob us."

I told him, "They are cowards, they only come in the night."

He looked at me and said, "Yes, but the improvised explosive devices work in the day."

I thought about that and was scared again. But that day, the road was clear and we arrived safely. I didn't know where I was going from Camp Phoenix, only that it would be to help the American military. September 24, 2010, was the first time in my life I had left Kunar Province. I wondered if I would ever see the mountains again.

4

Embrace the Suck

TOM

IN ALL THE things I do, I am driven to try to create a different life and trajectory as an adult than what I experienced as a child. As a result, I could not wait to move to Loyola University–Chicago, on the north side of the city, an area the opposite of my neighborhood in ways far more than geographic. As part of that move, I was excited to attend NROTC orientation the summer before my freshman year. Since I had barely learned of the existence of NROTC in time to join, I had only vague notions of what was going to happen at orientation. I expected some transformative challenge. I wanted a demarcation indicating that I was on the path I sought—that I'd left chaos and destinies written by surprise pregnancies and substance abuse behind, in favor of order and forward progress. I wanted a clear signal I was on a new path as part of some kind of meaningful service. I needed it to be clear. I did not know how to chart the path myself, and I definitely did not know how I would recognize it when I'd finally arrived.

What I got instead was disappointing: a little running, a little marching, a little yelling. It's not as if I knew what I wanted. I just knew that the vision of the Navy presented by NROTC was not the transformative moment I sought. I was six foot two and 140 pounds of debate club captain, a bench-riding, bookish kid whose mother described him as "the most accident-prone human I know." But my competitiveness drove me to seek the greatest challenge and the highest standards, and I was underwhelmed by everyone and everything at NROTC orientation. Whatever I was looking for wasn't there. Except the Marines.

There was an ascetic aspect to the Marines that appealed to me. They seemed to be a rejection of the male examples I was running from, men who relied on anger, intimidation, and physical force to make up for their own deficiencies and inabilities to effectively communicate. The Marines never stopped talking about attention to detail. They projected self-control and restraint while being forceful and aggressive. Physically, the Marines had a uniform hardness and physical fitness the Navy people just did not seem to possess. They were clearly more disciplined, more demanding. "Suck it up, harden up, and run faster" was their answer to most problems that week. The Marines identified weakness quickly, though not always fairly, since anything that was done incorrectly seemed to indicate great personal weakness in the violator.

Little in the way I lived my life, beyond some hard labor at my father's house and a busy high school schedule, indicated I was looking for what the Marines had. It certainly wasn't a conscious desire. I suspect I had a hidden belief that the hardest form of service must be the most authentic. In my eighteen-year-old mind I did not know what it meant to be a Marine, much less understand what it would demand of me. I just looked at the Marines on staff and the NROTC Midshipmen who were about to be commissioned into the Marine Corps and thought, "They have their shit together. I want to be more like them." They gave off a light in which I wanted to see myself reflected.

I am a guy who figures things out by doing them. Suddenly "Navy lawyer" was no longer my default aspiration. By the end of a week at Great Lakes, I wanted what the Marines seemed to have, even if I did not know why or what that really meant for me or about me. The Corps believes there is value in first appearances. Certainly my reaction to my first Marines would indicate that's the truth.

Pursuing the Marine Corps *was* a weird instinct for me because now I generally knew *what* I wanted to do, but I did not have a corresponding *why*. On 9/11, I had an urge to somehow respond, but I had no concept of what that might mean. I had a general desire to be of service to the nation and I had a corresponding need to "do something" specific to that day, but I had no overwhelming drive to fight as part of that service. I really had no drive to do anything specific. I just wanted to be of service to America in a way that also helped me move to a new place in life. It was a nebulous desire, but no less genuine for lack of definition. I just had no one to explain what options might exist for service beyond Chicago PD and Chicago FD, or what the choices I might make could mean.

My mother certainly exemplified service, both to her family, by taking a dangerous job she did not really want to do because it gave us things we needed, and to community, by simple virtue of doing that job. But since she didn't encourage me to join the military, I had no one else to model any martial values for me. Like anyone in 2004, I saw the news coming out of Afghanistan and Iraq, but I was certainly never one of those guys who watched a bunch of war movies. I didn't fantasize about winning glory through some display of martial valor or making a last, desperate stand on some desolate hilltop. Growing up, we were too busy actually being desperate to fantasize about it.

FOR ALL MY attraction to the discipline and "find a way or make one" spirit of the Marine Corps, I still had some growing up to do. I had pushed myself relentlessly so I could get into college. But I failed to understand that college was not the goal itself. I had no one to mentor me any differently. No one in my family or social circle went to college. I had no one to say, "Hey, get good grades because you need them to get a good internship so you can get into a top-five MBA program." I knew that college was something where people who aren't poor go, so I assumed that if I went to college, then magically, I wouldn't be poor anymore. To me, that college diploma meant people pay you a lot of money to do a job because you have the diploma. That perception was a working-class trap my wealthier peers never risked falling into. They had people in their lives to explain that college is a means, not an end. I didn't even know enough to be intimidated, so I just floated along.

Compounding the issue, my NROTC unit was located at Northwestern, a forty-five-minute trip on the "L" Red Line with a transfer to the Purple Line from Loyola. I had less daily military supervision than my fellow NROTC Midshipmen attending Northwestern. I was impressed by the Marines I saw at orientation, but there is a difference in talking about something and being about it. I was doing enough to get by academically, and was at best in the middle of the pack physically.

By the end of my freshman year, I had achieved neither the academic nor physical standards required for the NROTC scholarship I had dangled in front of my mom to overcome her objections the summer before. She had already dug herself deeper in debt to get me through the first year of a private university. If I had been more mature, I would have recognized that fact by working as hard as I could to make the kind of grades that reflected her sacrifice. But I didn't. I was too busy just getting by.

By the midpoint of my sophomore year, I still had not gotten a scholarship and it was not looking promising. Without a scholarship, or at least selection at a board as a "Marine Option NROTC Midshipman" I was tracking toward a commission in the Navy. I was also getting all of the additional pain associated with being an aspiring Marine in NROTC, with few of the benefits. Physical training, land navigation training, and conditioning hikes all ate up time. Additionally, like all potential Marine Officers in the Northwestern/Loyola combined NROTC program, I was required to be a member of the Rifle Drill Team. Several times a week, we met and practiced the manual of arms, a series of commands and responses that goes all the way back to the days of military formations lining up and shooting at one another across open ground. Truthfully, I've just never really been interested in, or naturally good at, things that require the crisp execution of military drill and I was about as good at the manual of arms as I had been at football. Nonetheless, I spent forty-five minutes each way to get to Northwestern from Loyola so I could stand or march while responding to the same commands, over and over, for hours. It was mind numbing. And I still sucked at it after two years. It drove our Assistant Marine Officer Instructor, Gunnery Sergeant Christopher S. Steele, the enlisted Marine charged with producing his own future leaders, crazy.

"Schueman, slap that rifle like you mean it! SNAP and POP!"

"AYE-AYE, GUNNERY SERGEANT!"

"Schueman, son, why do you suck so bad at this?"

"NO EXCUSE, GUNNERY SERGEANT!"

"Schueman, I give up. You ain't ever gonna learn to march. You are now the drill team photographer."

"GOOD TO GO, GUNNERY SERGEANT!"

I've never been so happy to be fired.

On top of drill, I had to be at Northwestern at 5:30 A.M. during the week to work out with the other prospective Marines. They were brutal, "run till you puke, do calisthenics till you drop" kind of events. Gunny Steele was notorious for finding random heavy objects, few of them ergonomically designed for runners, and sending us running up Sheridan Road along Lake Michigan with them. We were always watching for him out of the corner of our eyes. His inability to surprise me at Loyola was one of the good things about the distance from Northwestern. But when I was there, and he found me, he gave me "opportunities to excel."

"C'mere, Schueman!"

"Yes, Gunnery Sergeant!"

"Hey, pick up that sandbag right there! You know what to do."

"Aye, aye, Gunny!"

And off I went.

The wind off Lake Michigan was like knives made of ice flaying my face. I had been impressed at NROTC orientation by the physiques methods like his seemed to produce, but I was still tall and skinny and my physical fitness scores were just average for a guy who wanted to lead Marines. It got me attention I did not want.

"C'mere, Schueman!"

"Yes, Gunnery Sergeant!"

"You so tall. Must be good to be so tall. Ain't it good, Schueman?"

"Yes, Gunny!"

"You so tall, I bet you could reach up and pull down the sky if you weren't so weak!"

"Probably so, Gunny!"

"Get on that pull-up bar, son. Practice pulling down the sky."

I did pull-ups. Usually until he tired of counting.

"No, Schueman! That one don't count! That's twelve!"

I continued pulling myself over the bar.

"Still twelve, Schueman! Twelve! Twelve! Oh, there you go, that's thirteen!"

I was failing to embrace the suck.

Without a scholarship, I had to work just to live, never mind make tuition. I got a job at Costco, the membership-based bulk product dealer. Eventually I got assigned to the Costco tire shop. It was the best job I've ever had. Every weekend, I was plugging, mounting, and balancing tires. It was good, cathartic, purposeful work that made people's lives better, exactly what I've always sought. I thought back to my father's gravel pile in the barn. I stayed at Costco till my senior year, when they put me on shopping cart duty as punishment for plugging a buddy's tire for free on my lunch break. It was the middle of winter in Chicago and I had a career starter loan to cover my needs. I took military leave and never went back. I'm still officially on military leave at Costco, so I guess I can go back to work there someday, though I would probably have to start at shopping cart duty. Honestly, I would love to work there again. It was important work that made me feel good.

Work was not a big deal anyway; I had been working since I started high school. Then I worked every night from five to nine setting up life insurance physicals. When someone buys life insurance, they require specific physical exams. Usually, the people I called had forgotten they bought the policy requiring the physical I was calling about. As soon as I started the explanation for why I was calling, they assumed I was a telemarketer and hung up. I had fifteen seconds to convince them that they needed the thing that they had asked for in the first place and in fact had already paid for. My manager had a deal that if anyone signed up seventy-five physicals in a night, they got a pizza. I was so successful they bumped it to one hundred. That was actually a problem.

The ability to make a pitch in fifteen seconds helped me with meeting women. I'm not a "dude" guy and I am not much into "guy things." I am not passionate about cars or sports or scotch, though I am happy enough when our paths intersect. I didn't even start drinking beer until college, because I saw alcohol and drugs destroy too many people in my family. Cigarettes were out ever since I was nine years old and had to call 911 and direct them to my paternal grandma's house turn by turn because her oxygen machine had cut off and emphysema was strangling her in front of me. I only had one real vice. I desperately needed validation from women, and I was pretty reckless in seeking it.

THE SUMMER I was fourteen, I had gone to an evangelical Christian camp at Jekyll Island, Georgia. I didn't know much about God beyond the Our Father and the Charlie Brown Christmas story but there I had a very powerful encounter with the Lord. We were singing and praying at the camp and I ended up on my knees accepting Jesus Christ as my personal savior. I was born again. To me, there's my life before confessing my sins, accepting forgiveness, and believing that Jesus Christ is the Savior, and after.

When I came back to Chicago, I was really on fire. I was watching the religious channels on TV, praying five times a day, and giving my testimony. I told my mom, "We're going to find a church." That led us to a nondenominational church. I joined the youth group there and met a girl named Andrea Olivo.

Andrea and I had our first date collecting canned goods for the church youth group in which we met. We held hands for the first time on October 31, 2001. With two parents who clearly loved one another and welcomed me into their stable home, the Olivo house was everything mine wasn't. Her mom was sweet, her dad soft-spoken and mannered. He wore a button-down shirt tucked in with a belt. That was not something you saw where I grew up. Their home was a sanctuary. People ate at the table, together, at the same time every day. There was no yelling. They welcomed me and gave me a place to escape when I needed it. It was everything I wanted and nothing I knew. At their house, I felt safe. I felt stable. I felt normal. But sometimes I felt like I was looking through a window at a display in a store I could never enter.

Four years later, Andrea and I were still dating and she was still at the center of whatever stability existed in my life. But I was at Loyola, she was at DePaul, and I was horrible to her the whole time. I was never faithful. We would date, I would cheat, we would break up. Wash, rinse, repeat. After therapy, the origins of this are not particularly challenging psychology-wise. You learn things growing up in a broken home. Those things either become your normal or the thing you're running from. For me it was both. Having grown up in chaos and disorder, I was comfortable there. Because anything could change at any time in my childhood, whenever my life started getting too stable, too ordered, I sought a way to introduce chaos to make it feel normal. It affected my relationships. It affected my academic and military pursuits. It was compulsive, destructive, and counter to everything I claimed as my religious beliefs.

BY THE END of my sophomore year, I was increasingly familiar with the notion that "if the minimum were not good enough, it wouldn't be called the minimum." That's not really the attitude the Marine Corps seeks to imbue in its officer corps, but it's where I was, despite the conflict with my attraction to the higher standards of the Corps. I had gotten good at compartmentalizing. It made it easier to profess one set of ideals and live another. So just as I did with Andrea and my approach to our relationship, I claimed to believe in the discipline, fitness, and unbending standards of the Corps, but put good times in the moment ahead of all other efforts. Eventually Gunny Steele pulled me aside and said, "Schueman, with your grades you probably ain't going to get an NROTC scholarship. If you want to be a Marine Officer, you need to look at other options." It was a decisive moment. I had to take a hard look in the mirror and ask whether I really wanted a Marine commission badly enough to do what I had to do.

The Marine Corps has more than one path to an officer's commission. Attendance at a service academy or NROTC are the most publicly known. But there are also several programs that get you to Officer Candidates School (OCS). One is called the Platoon Leaders Class (PLC). PLC is split into two six-week sessions called Juniors and Seniors, executed over the course of two consecutive summers. PLC Juniors is largely a six-week physical grind designed to weed out people who will opt out of returning for the second miserable summer, called PLC Seniors. The second summer is an evaluation to determine the viability of candidates for training as lieutenants at the six-month Basic School.

I knew attending PLC Juniors certainly would demonstrate my seriousness about the Corps to an NROTC scholarship selection board. If the scholarship didn't work out, it also paved the way for attendance at a second summer the next year, which, if I passed, was another means of commissioning as a lieutenant. In the summer of 2006, I went to PLC Juniors and completed the six weeks of running, crawling, obstacle courses, and, of course, marching and the manual of arms. As we were practicing for our graduation parade, Gunny Steele, serving as an instructor over the summer, slipped up to me and said, "Well, Schueman, you just wasted six weeks. You had a scholarship for your junior and senior year before you even came. I just didn't tell you."

I'd spent six weeks with my head shaved bald, being screamed at for doing things incorrectly that I didn't want to do in the first place. I'd crawled through fetid swamps and run countless miles to earn a scholarship I already had, but it wasn't time wasted. I had been presented with a hard choice that involved accepting discomfort and adversity in the form of extreme discipline and arduous physical activity. It was an affirmative decision to work hard to earn what I ultimately wanted. I exceeded the minimum and it felt good. I started to make some mental connections about the value of working hard toward specific goals that summer. I realized that the active choice I had made to achieve a Marine commission made the goal that much more valuable. That realization was a lightbulb moment that clarified for me that I was not a natural talent, that I had to put my head down and grind to succeed in the face of adversity and if I got an opportunity to take some extra time to prepare, I should. I don't have much talent, but I can do work. That lesson about tough love, hard work, and the importance of standards is a gift given me by the man who would later be Master Gunnery Sergeant Christopher Steele.

NROTC scholarship students also attend Officer Candidates School for a single six weeks after their Junior year, so I still had to go back for six more weeks of embracing the suck. But I returned with the beginnings of a more stoic outlook, one that acknowledged, though did not yet fully embrace, that the severest school would be my future.

IN MY SENIOR year, with a scholarship, I was over the hump money-wise, but there was a new problem. My metabolism slowed down and my regular diet of pizza and beer suddenly started yielding different results on my perennially skinny frame. I wasn't prepared for the metabolic cost of that in the "appearance is reality" Marine Corps, where fitness is a virtue superseding most flaws.

In the Corps, fatness is an unpardonable sin. Gunny Steele again took me on as a personal mission. He was the picture-perfect Marine. Thin. Hair cut high and tight. Multiple meritorious promotions because he was a genuinely squared-away Marine. I still kept my eyes peeled for him whenever I was at Northwestern, trying to avoid him and heavy objects. But one day he called me on my cell phone. There was no avoiding him. He ordered me to come to his office. That was never good. This was no different.

"Schueman, you might let yourself become a fat piece of shit, but I ain't. We ain't commissioning a pig to lead Marines."

Normally I would have at least offered an "Aye, aye, Gunny!" but I was stunned into silence. He continued.

"You are gonna come in here every morning for PT. After that, you are gonna have a uniform inspection. Every day! Every day of your life is going to be the worst day of your life until you conform to standards."

Every morning, the PT was brutal. I lost track of the times I did push-ups and stair climbers until I collapsed in a puddle of sweat. Once I reached a point at which thrashing me further was pointless, Gunny would tell me to stand at attention. "Okay, Schueman, you got ten minutes to change over and I'm doing your inspection." He inspected me in a different uniform every day: Navy White Dress Uniform, Navy Service White Uniform, Service Dress Blues, camouflage utility uniforms. Passing was virtually impossible when I was still pouring sweat an hour later.

After several weeks of it, my best friend, J.R., was in my room talking to me as I neared a nervous breakdown. I'm terrible at asking for help, but I was cracking under the relentless torrent of Gunny's attention. With my head in my hands, I said, "J.R., I need help. I don't know how I'm going to get through this." He said what best friends say: "Okay, I'll help, get your uniform ready." J.R. modeled the kind of leader I wanted to be.

I thought back to the gunny we had during my first two years of NROTC, Gunnery Sergeant Joel Collins, later a sergeant major, the highest rank attainable for enlisted Marines. Gunny Collins was my first impression of a Marine and his is one that has always lasted with me. Every morning at 5 A.M. he drove a van from Northwestern to Loyola to pick us up and take us to Northwestern for drill or physical training. It was so much better than standing at the Red Line waiting, then connecting to the Purple Line to get out there. He was a guy who had kids of his own. He could have not done what he was doing and no one would have ever questioned it. I remember thinking, "I don't think this guy has to do this. I don't think it's in his job description that he has to go pick up the NROTC students at Loyola." But he cared. Over time, I definitely came to understand that Joel Collins was truly looking out for our welfare. He was living the Marine Corps motto, *Semper Fidelis*, "Always Faithful." That you can be relentless in service to Marines was an idea that came more fully to me later, but Joel Collins planted the seed.

I don't have many regrets in life, but one is that I squandered my time in college. I wasted an incredible opportunity to get smart, and to get fit, and to be a contributing member of society. Instead, I was a loud-mouth punk college kid who thought I knew everything even though I was an idiot. Regardless, it came together thanks to kindness and meanness alike from Marine Corps gunnery sergeants. I credit both Gunny Collins and Gunny Steele with teaching me valuable lessons about leadership that I still use today. With their attention, I graduated on time and got commissioned in the spring of 2008. I was headed to The Basic School (TBS) at Quantico, Virginia, the six-month entry-level course that all Marine lieutenants attend on entry to the Corps. TBS was where I would seek and earn assignment in one of the Corps' twenty or so officer's career fields. I had no clear vision of what specialty I would seek as a second lieutenant of Marines, but I was very proud to be one.

One other person maintained her standards. When I had my com-missioning ceremony, the NROTC staff asked parents to come up and pin lieutenant's bars on their Marine's Dress Blues. My mom refused. My grandma and my dad did it. Mom said she wasn't going to be a part of sacrificing her son.

FORTUNATELY, I HAD LASIK surgery on my eyes scheduled im-mediately after commissioning. That gave me a five-month delay before reporting to The Basic School, time I used to join a CrossFit gym and get fit to fight.

Because of the time of year that I checked in to TBS after the sur-gery delay, I was assigned to a Basic School training company half full of future pilots and staff judge advocates, Marine lawyers like I had told my mom I planned to be in the Navy. The other half of the class, including me, had to figure out what we wanted to do. All Marine enlisted personnel, regardless of specialty, are considered to be riflemen, and The Basic School is designed to teach all Marine lieutenants how to be provisional infantry platoon commanders. But its other purpose is ranking and evaluating all of the lieutenants in a class in order to assign them to various Military Occupational Specialties. We were all competing for a Military Occupational Specialty (MOS), the career fields into which the Corps apportions its officers. Once that apportioning is complete, lieutenants head to follow-on training in the areas in which they will specialize.

There's a common conversation the first few weeks of The Basic School that revolves around lieutenant aspirations. The pilots talked about what aircraft they hoped to fly. The lawyers talked about desirable duty stations and practice areas. That left the rest of us to have conversations that were mostly empty male posturing. In the first few weeks of TBS, most of the class claimed to want service in the ground combat arms, preferably in infantry, with tanks or artillery as bridesmaid choices.

"Hey, man, what MOS do you want?"

"Infantry. That's the real Marine Corps, bro."

For my part, I was honest—I told people I had no idea. But by the time we completed our first field exercise, the burning embers of my commitment to the Corps caught flame. I had finally found the purpose hinted at in my initial exposure to Marines. The exercise was called Field Exercise 1 and was essentially spent walking around with guns in the dark woods, looking for other platoons of lieutenants so we could shoot at them with blanks. To anyone who had their eyes on the clouds and a future in a fighter jet, it could only have felt like marking time. But I loved it.

FEX 1 left me exhausted, but my first taste of infantry life awakened something latent in my unnatural competitiveness, an aggression and a predatory spirit that made the hunting of armed men seem like an aspiration. I had always wanted to win. The infantry-focused training at TBS taught us that doing whatever it took to do so, within the standards of discipline and the laws of war, was the standard. In the infantry, that desire could be distilled to its purest essence: kill or be killed. Because my childhood left me comfortable with chaos as a default state, now, in tactical exercises, I could see the way scenarios would develop before they did. I knew by instinct where the opposing force of lieutenants was arrayed relative to my own, even when I could only see a few Marines to my left and right.

Many of the class emerged from that freezing week spent walking and (barely) sleeping in the woods with a suddenly keen interest in administration or aviation maintenance or anything that didn't involve walking. Their rejection of the spartan life of the infantry made it all the more appealing to me. For the first time in my life, I knew precisely what I wanted to do and why. I wanted to be an infantry officer; a grunt. That certainty was the most exhilarating and satisfying feeling I had known in my life.

Initially I thought I wanted to be a grunt simply because I didn't want to be with the people who didn't want to be grunts. It was more a rejection of something than an embrace. But I increasingly realized that I was twenty-two years old and had found my purpose in the woods of Quantico, Virginia. I heard music in the thunder of weapons firing on a range and found reward in hand-to-hand combat, where the aggression that fueled my unwillingness to lose was no longer an aberration, but an asset. My decision to seek an infantry assignment was an acceptance of who I am and a pursuit of who I wanted to be. I was finally operating from desire rather than desperation. Unlike in high school, when I had achieved for achievement's sake, unlike my four-year drift through college, now I had purpose.

I wanted acceptance into the infantry, the beating heart of the Corps. That meant being harder and faster and more aggressive than my peers. I was fortunate that a lot of them were already bound for flight school in Pensacola, Florida, or Naval Justice School in Newport, Rhode Island. They had already opted out of a life of fighting holes, maximum effective ranges, and long hikes under load. But there were still enough aspirants that I had to compete for what I wanted.

I worked hard at The Basic School and I was rewarded with the infantry assignment I wanted. As long as I passed the Combat Endurance Test, I was headed to the Infantry Officer Course (IOC), widely seen as one of the toughest schools in the Corps and the entire Department of Defense. The test was brutal enough that it was later removed from the list of requirements for entry to IOC in 2018. But it was 2009 and the Corps was in the middle of two hot wars. Ensuring that IOC was producing the leaders that young Marines needed was of critical importance. So, we would suffer. I was certainly faster and stronger than I had been. But in my head? The guy who was slower and weaker than his peers, the kid dressed out and on the bench at Marist High School, was whispering. I had to defeat him.

Fortunately, just like that first six weeks at PLC Juniors, just like the postsurgery five months I had to get ready for TBS, fate and a training schedule dictated I had almost two extra months to prep for IOC. Every morning we spent hours throwing each other to the ground and practicing crushing an enemy's face with our boot heels as part of the Marine Corps Martial Arts Program. From there we ran the Endurance Course, a six-mile obstacle course made of trails winding up hills, over water, and down cliffs. We spent the afternoon in the gym, building physical mass that could take carrying heavy loads over many miles and hours.

Still, the Combat Endurance Test loomed.

The test involved incredibly long land navigation legs with map and compass. We ran multiple obstacle courses for time. There were physical fitness tests with events we'd never been subjected to, foreign weapons assembly and disassembly tests, technical and tactical evaluations, and swim tests in the pool. I'd never been a strong swimmer and swallowed a lot of chlorinated water. Add weapons, gear, and Quantico, Virginia's heat to all of it and the Combat Endurance Test was soul sucking. But I made it. I was going to IOC. Maybe there's a questionable value in competing for prizes that few people would want, but I felt validated. I won. I felt proud. I felt at home in a place that was made for people like me.

Every time I visited my real home, it was a fight over the Marine Corps. When I was selected for the infantry, my mom lost it. I remember her yelling, "The hell you are! That's not fucking happening! You'll be used as cannon fodder!" She mentioned something about shooting me in the kneecaps. She swore this was just "some macho shit" because she was a cop. She said it wasn't just me I was dragging down this road;

I was forcing her to travel it, too. But a young man has to live his own life and I had realized that mine was in the Marine infantry.

IN THE INFANTRY Officer Course building, a brick box called Mitchell Hall, I looked at the pictures of Medal of Honor recipients. Staring at the pack weighing down the Marine in front of me as we struggled up another hill, I thought, "I'm here. I'm home. But I have to earn the right to be here every day." Our IOC instructors were fresh off the battlefields in Afghanistan and Iraq. I wanted what they had, an air of confidence and solemnity. They were tested, and on the other side of it they were the kind of men and Marines I wanted to emulate badly enough to put in the work to do so. I was entranced when they spoke of their experiences at war. Combat seemed like an environment in which life would be stripped to its most elemental, a way to live a pure existence focused only on things that matter. I had a purpose. I wanted to go to combat.

At IOC, the infantry captain who oversaw our class told us, "The most important grade you will get here is the reputation you carry forward to the operating forces. Get hard and work hard, gentlemen." My classmates and I lived in the woods and desert for three months, sleeping under the sky when we slept at all. We were semi-savages, constrained only by the unforgiving standards of the Marine Corps. They were rules tied more to our future survival and that of our future Marines than to any normal societal mores, though our ethical discipline was expected to be inviolate.

We were preparing for a present war and were accordingly granted the professional freedom to try anything in training as long as it met minimum safety standards and was technically and tactically defensible. We shot every weapon in the armory, from pistols to machine guns to mortars and antitank rockets. We became expert in land navigation, in calling for indirect and aviation fire support, in raids and ambushes and movement to contact with the enemy. As preparation for the wars waiting for us in Iraq and Afghanistan, we worked our way through mock villages, engaging with role players who invited us in for tea, presented cultural conundrums, and shot blank weapons at us. We practiced responding to improvised explosive devices and the traumatic injuries that came in their wake. Mostly we spent weeks carrying the majority of our body weight in packs on our backs through the Virginia woods and California desert. Our hard-earned gym muscles got replaced by lean mass and a never-quit mindset. We learned to seek contact with an enemy and then press the fight as violently and aggressively as possible. I loved it.

Adding to IOC's allure, for a kid who once lay prostrate with grief over the absence of his father, for a bench rider who left the football team for academics when it became clear I was destined to watch from the sidelines, I was in the company of men who in some way fit the masculine ideal I had always sought, and they accepted me within the same paradigm. My peers were uniformly physically fit, tough young men, all looking to be tested for different reasons but willing to locate, close with, and destroy an enemy.

Infantry Officer Course 4-09 was made up of men like Ty Anthony, Steve Tryzna, Alex Pearson, Vince Young, Joe Szostak, Cameron West, and William Donnelly IV. Cam was an aspiring cowboy who grew up in the Atlanta, Georgia, suburbs, and was fast becoming one of my best friends. I was a city kid from Chicago, but Cam and I lifted weights and drank beer and rode around in his old blue Chevy pick-up truck. Will Donnelly, from Picayune, Mississippi, was the kind of guy who gets married on 9/11 and flies an American flag outside his house. In a crowd of wild men he was stable and calm and always smiling.

There were older lieutenants with enlisted service, like Johnny Eppes, Chris Kakas, Joe Patterson, and especially Robert Kelly. Rob was a general's kid with an older brother already in the Corps. He was smart and dryly funny, with a Bostonian's cutting humor. He didn't look the part of a Marine. Rob was skinny and always looked dirty in the field, though his uniform was always Marine Corps squared away. The Marine Corps is an appearance-focused organization where muscular, square-jawed guys have an advantage on day one. That was Cam: a big Georgia boy able to immediately relate to anyone with his charisma and charm. I had finally discovered weights and was filling out my six-foot-two frame, too. That wasn't Rob Kelly at all. He got by on performance and dedication.

Cam and I gravitated to him like an older brother. Rob's spare physique and nasal Boston accent, not the tough guy from Southie accent but the MIT accent, led Cam and me to spend much of IOC teasing him about taking up his evenings playing *World of Warcraft*. We often threatened to stuff him in a locker. But we truly looked up to him and that made it easy for Rob to use his wicked intelligence to cut us both in half with a joke. If the infantry is a dog pack where you have to fight for position, guys like Cam and I were puppies, excited and loud and always making noise. Rob was a greyhound with a German shepherd's heart. When we got too enthusiastic about teasing him, he bit back.

Coming back from the gym and finding Rob reading a book, and ignoring the fact he had probably already finished a run and been to the gym before we were awake, Cam and I would call him out, "Hey nerd, you missed today's lift sitting here while we were getting after it."

Rob would respond with something like, "Oh, getting after it like . . . you're fighting insurgents block to block in Fallujah for a month? Or getting after it like . . . you were lifting inanimate objects in an airconditioned room?"

It was as different a group of guys as you're likely to find in the Marines, but the more time I spent with them, the more I understood what I was doing there. In literature there is an archetype called the hero's journey. The hero's journey starts with a call to adventure. In retrospect, that was 9/11; I just didn't have the frame of reference to recognize it. IOC was the acceptance of the call I failed to recognize eight years before and the inception of the journey in a way that none of my previous military experiences had provided. At TBS and IOC I found within myself, and was validated for having, an intense aggression I didn't know I had. Somewhere in the woods of Northern Virginia I finally embraced the suck. I accepted, and reveled in, misery; in living by standards we knew would eventually mean life or death for some of us.

We accepted it all as part of the price of admission to an elite society. At IOC, I was experiencing violence for the first time. It clicked that I was now part of this violent thing, that there were two wars ongoing, and I was going to go do violent things on behalf of my country. I finally felt like, "Yeah, this is it. I didn't know what I was looking for up until now."

The heart-and-soul action elements of the Marine Corps, around which all other capabilities are built, are its infantry regiments and their three subordinate battalions. Those units are primarily located in three places: Camp Lejeune in Jacksonville, North Carolina; Camp Pendleton and 29 Palms, California, reaching from the beaches of Oceanside to the southern Mojave Desert; and Okinawa, Japan.

For a Chicago boy who liked to party, rural North Carolina held little allure. All the infantry battalions in Okinawa are really just forward-deployed units from Camp Lejeune or California, part of a shell game that puts battalions in the Pacific for a quick response to problems there. I could not be assigned there initially, though I knew I might deploy there in the future. Although units at 29 Palms are generally known to be some of the most technically and tactically proficient in the Corps, people ascribe that to the fact that there is nothing about life in 29 Palms to distract Marines from training.

The good life at Camp Pendleton was calling me when I graduated the Infantry Officer Course in 2009 and I asked for assignment to the west coast. When my orders came to join the Third Battalion of the Fifth Marine Regiment, shorthanded to "3/5" and pronounced as "Three-Five," I was psyched. So far, the Marine Corps had given me everything I never knew I wanted.

5

"If They Get Killed Here, Why Should I Not?"

ZAK

AN AFGHAN MAN from Mission Essential Personnel was waiting for us when we arrived at Camp Phoenix in Kabul after the long bus ride from Jalalabad. I was tired not just from the journey, but also from the stress of worrying about Taliban attacks. Though they had shot at all of us in Nangalam, and even though my truck had been the one to find an IED, as an interpreter I was now personally a target. The man waiting for us wore a black leather jacket and American baseball cap. He had a *dest-mal*, a wrap Afghan men use to keep dust out of their noses and mouths, around his neck. As we got off the buses and found our bags, he waved us over to him and said, "Follow me, we are going to a place where you can rest until we leave tonight." He led us to some long wooden buildings with bunks inside. It was still warm in Kabul and there was one air conditioner struggling to cool the room. The buildings looked the same as FOB Blessing in the Korengal, but no one was shooting at us, so I was happy enough. We sat down to rest and prepare for the next move. Before he left us, he called out to the entire group, "We won't be here long. Be ready to move." He pointed at a pile of bottled water and several cases of fruit and told us to help ourselves. "When it is dark, we will get on the buses to Bagram. We will fly from Bagram to Kandahar tonight."

I set my bag on a bunk and sat down. I leaned back against the bag as I thought about where I might be going. Kandahar was and still is the home of the Taliban. There was a lot of fighting in the Arghandab River valley and in Kandahar City as well. If I went there, it could be very dangerous. But Kandahar Airfield supported other bases. Going there did not mean I would stay there. From Kandahar, I could go to Herat, in the west, near the Iranian border. Herat was less dangerous than Kandahar, so that would be okay. I hoped that would be the place. But I read the news. I knew Helmand Province was most likely the place where I would go.

Helmand was originally part of Kandahar Province. The people there are just as connected to the Taliban as those in Kandahar. I thought about the Highlanders in Nangalam and how they tried to kill us for being outsiders. I knew that people in some parts of Helmand were like that. I also knew Afghanistan produces most of the world's opium and most of it comes from Helmand. The Taliban sold opium to pay for their war, so they opposed the government's attempts to stop farmers from growing it. I felt nervous. I lay down on my cot and tried to sleep until the buses came. I did not sleep well. I don't know how long I lay on the bunk with my eyes closed until the same man as before came inside and told us, "Get your things and go outside. The buses are here, we are going to Bagram."

BAGRAM AIRFIELD WAS the largest military base in Afghanistan. It is in Parwan Province, about forty miles north of Kabul. The Russians had Bagram when they were in the country. Then the Taliban and the Northern Alliance occupied different sides of the airstrip and spent years fighting over it. There were still trenches from that time on the northern end of the airfield. When the Americans invaded in 2001, they took over Bagram and began developing it. By 2010, the perimeter was eight miles around. Security measures were high. The base was surrounded by twenty-foot cement blast walls to protect people from bomb blasts. Our buses had to wait for forty-five minutes to get through the security gates. Once we were inside the airfield, the buses took us straight to the airstrip. When we got off the buses, I was standing at an airport for the first time in my life.

It was nearly midnight so the sky was dark. But huge lights on the airstrip made it as bright as in the day. Planes of all size were parked on the side of the airstrip. The noise from the jet engines was overwhelming, louder than anything I had ever heard. I could not talk. I could barely think. As I stood in line to get on the plane, an American fighter jet roared down the runway. When it reached the end of the runway, it shot into the air trailing a massive flame from the exhaust. It seemed like we were consumed by one continuous roar as aircraft moved around, taking off and landing.

A uniformed American wearing a headset waved to us and turned, signaling us to follow him to the back of a huge, gray American plane. The line of Afghans walking in single file behind a uniformed American felt like a symbol of the bigger war. The airplane crew had us pile our bags on aluminum pallets on the floor as we stepped up onto the ramp and inside the plane for what was, for most of us, the first time in our lives. The American wearing the headset connected it to a box on the wall of the plane, then pointed at us and gestured where we should sit. As he moved, he kept adjusting the cable that connected him to the airplane. He was shouting into the mouthpiece, but I could not understand what he said over the engine noise. We did what we were told and sat along the walls of the plane in seats made of nylon straps. More Americans strapped nets over our bags and tied them down on the back ramp of the plane. The ramp began to close and our baggage rose with it. Soon we were sitting in the plane with a dim green light glowing overhead. From inside the plane, the engines were quieter, but the hydraulic system made a constant high-pitched scream. It was not like any experience I had ever had in my life.

The military plane was not like the civilian jets I had seen in pictures. There were no windows I could see out of, so when the plane started moving, it felt strange to me. I had never felt motion without being able to see proof that we were moving. In the future, I would be bored during aircraft flights except when they had problems and had to land, but this was the first time I had ever felt the sensation of moving this fast. I was kind of scared, kind of excited. With the angle of the aircraft as we lifted into the air, I could not help but lean into the person sitting beside and suddenly below me. I wanted to apologize but did not want to shout into a stranger's ear over the noise of the plane. Two hours later, we landed at Kandahar Airfield. It was the middle of the night, but like Bagram, there were people working everywhere, aircraft moving. That's how the next year of my life would be.

From the airstrip, we loaded into buses again. They took us to a special camp for interpreters, surrounded by a fence. By the time we were all gathered, there were about sixty of us staying there. Like at Camp Phoenix, we had large communal bunk areas. Like at FOB Wright and Camp Phoenix, an Afghan man explained to us what was next. He told us to go to sleep and that we would get more information at a meeting in the morning.

As I lay down to sleep, I thought about the meaning of what I was doing. Places like Kandahar were dangerous. I had heard some of the other translators talking. They did not want to get hurt. They were talking about what they would or would not do if the Americans told them to. I did not want to get hurt, either, but I thought, "Someone has to take a chance for Afghanistan." I was not in the Afghan National Army, but I felt that by using my skills to help people trying to help Afghanistan, I was serving my country and building our future.

I had seen the Taliban's Afghanistan and by 2010 they were coming back in places like Kandahar and Helmand. In the east, where I was from, they were coming back out of the valleys. I did not want to go back to a time when there was no free expression and no education. I did not want my sisters to grow up in a place where they could be beaten for not covering their faces. I was willing to fight for those things. I was willing to die if I had to. By going to war, I would save money while I helped my family. I still wanted to go to university, and translating for the Americans was a way to fund it. I knew I was making the right decision. But I still did not sleep well.

I was awake and had performed my ablutions and prayers well before the man from MEP came back. "Everyone come with me. I will take you to breakfast and then we will give you information about what is next." We all followed him to a special dining facility on our camp. It had Afghan food. Once everyone was there, the meeting began. The MEP representative told us that if we wanted the job as translators, we were all going to Camp Leatherneck in Helmand Province. Even though I knew I was doing what my family and my country needed, I was immediately scared. A few interpreters resigned on the spot when they found out. The British had been in Helmand since 2006. Their time there was very violent. Now, in 2010, the provincial capital, Lashkar Gah, was under threat from the Taliban. The Americans were taking over. There was no reason to expect anything different in Helmand now, unless Afghans like me helped the effort.

Helmand is a Pashtun province. I knew I could speak to the people there to try to help them understand what we needed to do to make Afghanistan better. But the people in Helmand have a reputation for not liking foreigners. I did not know if they would be happy to see a person from Kunar, even another Pashtun. All I could think about was the news I had read about the fighting in Helmand. Now I was going there. I never considered quitting, though. I thought back to a conversation I had with my father before I left our home in Kunar. He and my mother knew we needed the nine hundred dollars a month that MEP would pay me, but they were concerned about my safety.

"I want to help Afghans not get hurt in the fighting!" I told them. "I want to help the Americans help our country. They are here to help us and I will accept what they accept. If they will get killed here, why should I not?" I had to do what my country needed so we could have a chance at making a better place.

The United States had tried to help the people in Helmand before, from the 1940s to the 1970s. Helmand is the largest province in Afghanistan by size. Much of it is the Registan Desert, which is made of a fine powder that is easily disturbed by footsteps and gets in your nose, eyes, and mouth with a little wind. One point four million people live in Helmand and they must eat and drink, so many people live along the Helmand River, which is the major source of water for drinking and irrigation to the province. The Americans built a dam at Kajaki in 1953. In 1975, they installed hydroelectric turbines that provide most of the electricity to Helmand Province. The Taliban tried to destroy it in 2007.

The Helmand River feeds a system of canals and underground watercourses called *karez,* built to help the people irrigate their crops. The Taliban liked to store weapons and drugs in the *karez* and use them to move without being seen by aircraft. I did not know that sitting at Kandahar Airfield. I just knew that no matter how dangerous it was, there were ten people in my family who needed the money I was going to make.

After two weeks in a tent at Kandahar, we finally moved to Helmand Province. It was like the trip from Bagram to Kandahar. We got on buses to the airfield. We got on another big gray plane in the dark. We flew in the night and landed at Camp Leatherneck in Helmand. The British still there operated from their base, Camp Bastion. The Marines built Leatherneck as their base when they took control of Helmand Province from the British. The two bases were connected. Helmand was as different from Kandahar as Kandahar was from Kunar. When we came off the plane, I heard another interpreter say, "We are in the dirt! It's all dirt! No trees, no jungles, no mountains! Everything is just gray! No trees, no grass, nothing!" That was true. There were none of the cool rivers and green mountains of Kunar. We seemed to walk through miles of gray dust and gravel. I missed my family and my home already.

At Camp Leatherneck we had another special camp for interpreters but we ate at the American dining facility. The Marines called it a DFAC, so we did, too. A Marine escorted us from our camp to the DFAC three times a day. I missed the Afghan food at Kandahar. The American food at the DFAC was strange and bland. Americans did not seem to eat vegetables! Other than eating, we had little to do. We spent some time getting equipment: boots, uniforms like the ones the Marines wore, helmets, and body armor. The rest of the time we sat and talked about what might be coming next. Whatever our duties would be, there was no training for it, just waiting and wondering when we would go to the next place and what would happen when we got there. I was bored. But I thought about Nangalam and working behind screens to keep from being shot. I thought I was better off bored.

Finally, after three or four weeks at Camp Leatherneck, a Marine came and told us to be ready to leave the next day.

Someone shouted at him, "Where are we to go?!"

The Marine looked serious and said, "You all are going to Sangin District. Some of you are going to work with Third Battalion, Fifth Marines. They just took over from 3/7 and the British."

Helmand was one of the most dangerous provinces in Afghanistan. Sangin was the most dangerous district in Helmand. It was a key part of the opium business for the Taliban, so they fought for it. Also, they knew that Afghan and foreign troops were stronger because of airplanes and helicopters, so they protected it with hundreds of bombs. There did not seem to be an end to them. Afghan and foreign troops were killed or badly injured all the time. More interpreters resigned immediately.

THE NEXT MORNING, a Marine sergeant came to get us. He said, "Bring your bags and follow me. We're fuckin' movin'." English was hard enough to learn. A lot of the words were new to me. I was learning the Marines spoke another kind of English. A lot of it was the word *fuck*. Because I had carried bags for America soldiers in Asadabad, that word was not new to me, but these Marines said it all the time, for everything.

"Go ahead and put your fuckin' body armor and your fuckin' helmets on." We followed the sergeant to a line of vehicles. The Marines had huge, seven-ton, armored trucks. They had armored vehicles they called MRAPs. It stood for Mine-Resistant, Ambush-Protected. A Marine called us interpreters to gather around him along with the Marines. I learned later that he was a gunnery sergeant.

"Stand by for your fuckin' convoy brief." He had a map taped to cardboard. As he showed us the route we would take to the next place, FOB

Nolay in Sangin District, he pointed to places and told us what happened there and what we would do if it happened again.

"Fucking Brits got fucked up there. Keep your fucking eyes open for wires."

"When we get there, watch this fuckin' intersection. There was a complex ambush there. Brits had three KIA. If we take fire, you return fire and suppress so affected vehicles can get out of the fucking kill zone."

"Lima Company hit a fucking IED here three days ago. Went low order and fucked up a Marine's leg but he will be okay."

I did not know what everything he said meant, but I understood what I needed to understand. This was not going to be easy duty. But it was *my* duty. One interpreter must have understood all of it because he raised his hand and resigned. He walked back to the interpreter camp. I watched him go. I thought of Nangalam again. It did not seem as bad.

FOB NOLAY WAS a bigger base that supported smaller FOBs and observation posts (OP) in all of Sangin District. Our movement there was a safe one and I was again waiting for more directions. I was getting used to waiting for Marines to tell me what would happen next. I was more nervous at each new place because I knew that each move was a step closer to the war, but I also wanted to get to the end and find out exactly how I was going to help the Americans and my country. After one week in a tent, a Marine staff sergeant came to us and announced, "We're taking you to FOB Inkerman tonight. Terry Taliban likes to sleep in, so it is safer to move in the night."

I was not the only interpreter to make the move to FOB Inkerman. There were three of us who arrived there after midnight. With me came interpreters named Adan and Kamar. They were both from Paktika Province, in the east like me. The British built FOB Inkerman by putting up HESCO barriers and wire around old deserted mud compounds. In the dark, with only stars to light us, the camp was hard to see, but when we unloaded from the truck, I could tell Inkerman was not a nice place. It was like Topchi but ugly. The same Marine staff sergeant who had brought us from FOB Nolay told us to follow him. He took us into a room with lights powered by a generator making a lot of noise outside.

"All right, fellas, each of you pick a number."

I chose number one. Adan chose two. Kamar chose three.

"Okay, Zaki, you are going to First Platoon. They happen to be in from PB Fires today, so you can stay right here at FOB Inkerman tonight." I didn't know what PB Fires was but I didn't want to ask any questions then. He looked at the other two interpreters. He pointed at Adan and said, "You are headed to Outpost Transformer with Second Platoon." He turned to the other interpreter, Kamar: "You are going to Third Platoon at Patrol Base Fires, too. You can walk back with First Platoon and join them tomorrow. All of you, follow me." At least now I knew that PB Fires was another base for Marines.

The staff sergeant took us to a house made of mud bricks. He showed us a room, bare except for three bunks, a cardboard case of halal Meals, Ready to Eat (MREs), and a pile of sticks the original owners left. "This is your room. Help yourself to whatever bunk you want. I'll be back in the morning. If you got to piss, there are piss tubes outside. If you need to take a shit there is a WAG bag." I did not know what a WAG bag was, and I did not think I would like to find out. But I knew that whatever the Americans did, I had to do. The staff sergeant told me where to go to the bathroom and turned to leave. He stopped in the door and looked back at us. "Get some sleep, guys. You will meet your platoons tomorrow."

We each took a bunk we liked. It was October 13, 2010. The next day I was going to meet my team. First Platoon, Kilo Company, Third Battalion, Fifth Marine Regiment. Kilo/3/5. They were strangers and foreigners. Soon I would call them brothers. I knelt to pray before sleeping.

6

Darkhorse

TOM

DRIVING AWAY FROM Quantico felt like beginning a new chapter in my life. I was no longer a "student officer." I was a qualified infantry officer, a man with a profession, commencing its practice. As the miles stacked up behind me, I felt increasingly consequential. Certainly, I understood the stereotypes about inexperienced second lieutenants inadvertently making more problems than they solve, but I was excited to practice what I'd heard preached for a year.

I was not the only one of my friends making the same cross-country drive. Eight members of IOC Class 4-09 were also assigned to Third Battalion, Fifth Marines at Camp Pendleton, California: Robert Kelly, Cameron West, Will Donnelly, Joe Patterson, James Byler, Brad Fromm, Chris Kakas, and Luke Postma. As I drove, I thought about the previous year, a mental journey tied inextricably to 9/11. That Tuesday in September had not been the impetus for my joining the Marine Corps, but the last year of training and indoctrination had moved it, and Afghanistan, to the forefront of my mind.

Thinking about 9/11 reminded me of Brother Brennan and the morning that set all of the life that followed into motion. Brother Brennan was a member of an order to which he dedicated his life. I felt I understood him better now, that I too was a member of an order, one composed of warriors. We were conditioned to move to contact; to seek an enemy and draw fire. Afghanistan was calling us all. I wanted combat. Ergo, I wanted Afghanistan. I needed to know if I could succeed there, if I could be the kind of man I had come to so admire. Two members of our IOC class, Ty Anthony and Rob Fafinski, never made it back to Quantico for graduation. After completing the final phase of IOC at 29 Palms, California, they remained in California at Camp Pendleton and deployed to Afghanistan with Third Battalion, Fourth Marines a week later. I did not know what my assignment at 3/5 held for me, but I hoped I would follow them.

As I drove across the United States, I read the World War II classic *With the Old Breed*. The author, E. B. "Sledgehammer" Sledge, was a member of Kilo Company, 3/5 who fought at Guadalcanal, Peleliu, and Okinawa in World War II. I love the book, but it only tells part of the battalion's long, proud story.

Third Battalion, Fifth Marines is known in the Corps as "Darkhorse." A Darkhorse is a term for a smaller, little-known racehorse that delivers an outsize result. Lieutenant Colonel Robert Taplett was the battalion's commanding officer in the Korean War and used the call sign during the bloody, frozen fights of that conflict. The name went out of use for decades but the battalion asked to use it again during the early years of the war in Iraq to honor the battalion's combat history. It is an apt name for a unit for whom survival was in question on many occasions but who repeatedly turned doubt into victory.

Korea is hardly the only streamer that hangs from the 3/5 battle colors. 3/5 fought to victory at virtually every significant World War I fight for which Marines were present. In recognition of the unit's collective valor at charnel-house battles like Belleau Wood, forever known as Bois de la Brigade de Marine, the French government presented the Croix de Guerre to the entire battalion. A century later, 3/5 Marines still wear the French Fourragere, a cord looped around the left shoulder of their service dress uniforms, presented in recognition of valor a century before.

After World War I, 3/5 fought insurgents in Nicaragua. Counterinsurgency is grudge-based warfare. It's a back-alley knife fight writ large.

Combatants may know the names of the men they seek to kill. For six years, Darkhorse patrolled steaming jungles, trained local police forces, and attacked insurgent base camps while carrying lists of men marked for death. It was a taste of the one-by-one style of the bloodletting coming in the first part of the next century and the frustrations attendant to trying to change a culture under fire.

Soon 3/5 found itself in the South Pacific during World War II. In the jungles of Guadalcanal, New Britain, Peleliu, and Okinawa, the battalion was in some of the fiercest fighting in the Pacific Theater. Of Peleliu, Sledge wrote, "To the noncombatants and those on the periphery of action, the war meant only boredom or occasional excitement, but to those who entered the meat grinder itself the war was a netherworld of horror from which escape seemed less and less likely as casualties mounted and the fighting dragged on and on. Time had no meaning; life had no meaning. The fierce struggle for survival in the abyss of Peleliu had eroded the veneer of civilization and made savages of us all." It is a passage that could be used to describe much of Darkhorse's history. It is particularly apt in the "give no quarter" fighting that took place in the South Pacific.

By the close of World War II, the Marines and Sailors of 3/5 had endured a volume of sustained savagery previously unseen by the Corps. The unrelenting horror of that experience provided a base of valuable combat experience when the battalion, now officially recognized as Darkhorse, went to a "police action" five years later in Korea. Darkhorse fought battles now firmly fixed in the entirety of the Corps' legend, defying the odds at places such as the Pusan Perimeter, the amphibious landings at Inchon, and the house-to-house fighting in the streets of Seoul. During the breakout and fighting withdrawal from Chosin Reservoir, through encircling Chinese and North Korean troops, they gave life to the Marine Corps motto when they carried their dead and wounded out with them.

Thirteen years later, Darkhorse found itself in Vietnam, back in another grinding jungle war. Vietnam was an undeclared hybrid war demanding the sweat and tears of counterinsurgency and the blood of high-intensity warfare. For fighting on blood-soaked ground as diverse as Hue City and the Que Son Valley between 1966 and 1971, 3/5 twice received the Presidential Unit Citation. There is no higher collective honor an American military unit may be awarded but it is one that comes with a blood trail.

Since 1990, 3/5 has repeatedly seen action in the Middle East. During Desert Shield in 1990 and Desert Storm in 1991, the unit was in combat in Kuwait. Capable of offering a helping hand as much as swinging a closed fist, 3/5 responded to one of the deadliest recorded cyclones in history while en route home from Kuwait, providing critical humanitarian assistance in Bangladesh after much of the country was destroyed by wind and water.

In 2003, 3/5 was again in the Middle East for the invasion of Iraq, before returning to distinguish itself in November 2004 during the Second Battle of Fallujah in Al Anbar Province of Iraq. While preparing for that 2004 deployment, 3/5 began again officially using the call sign "Darkhorse" in communication. The notion of a small force delivering outsize results was appropriate. Fallujah was the most vicious city fight that Marines, or any American force, had experienced since Hue City in 1968. Darkhorse Marines were at both. Some of the members of the battalion I was joining had been in Fallujah in 2004. Many others had seen combat elsewhere. One of the few places Marines have been that Darkhorse had not was Afghanistan. In my warrior's heart, I hoped that was about to change.

I STOPPED AT home in Chicago on the way to California. My mom was even more unhappy at the direction I was taking in life than she had been when I joined NROTC five years before. She was still my mom nonetheless and decided to drive to California with me to help her baby boy get settled into his new life. Before we left, though, everything she had been holding in erupted: "Thomas! The thought of you not being able to make your own decisions, following orders to kill, it's a nightmare for me!" She stalked back and forth in our kitchen. I was not really in any mood to hear a lecture about the profession in which I had finally found my purpose.

"I'm finally doing the thing I am meant to do, Mom! God has a plan for me, a path of service. I was put on this earth to serve. It's what I want and this is how I'm supposed to do it."

She was trying to talk through tears. "Thomas! Do you not understand that if you die in some bullshit war, I will not survive it?! These people, with their, 'Thank you for your sacrifice'? I don't have a choice, do I? 'You must be so proud!' Yeah! I am proud! I've *always* been proud of you! Of your loving heart, your intelligence, your thoughtfulness. You don't need a uniform to show those!"

As a policewoman, my mother was charged to "serve and protect." She was a lifelong example to me about the essential requirement for selflessness as part of service. But her own service, aside from getting health insurance for my sister and me, was about providing a bulwark between chaos and order, and I believe she saw the military as sowing disorder. There was plenty of disorder in my family, but Mom kept the streets in the streets, where she saw her role as a peacemaker and peace enforcer. She viewed the military as a machine in which I, her "beautiful boy," would become a killer automaton. She had a very present respect for the military as an institution and she certainly approved of me serving in a nonmilitary capacity like the Chicago Fire Department. But she just wasn't willing to see me bear the cost of military service as she understood it. I think that like many parents, she saw her service as a bridge to something better for her kids, something where catching hepatitis wasn't a probability, much less gross disfigurement or death, and that meant we could serve without knowing those kinds of costs. Of course, as an infantryman, I accepted those costs as much as I could without actually having been in a position to risk them. That night she made me watch Tom Cruise as Ron Kovic in *Born on the Fourth of July*.

Before we left Chicago, in what is an almost ubiquitous Marine rite of passage, I got a tattoo, marking myself with Psalm 27:1: "The Lord is my light and my salvation; whom shall I fear?" It stands as a testament to my faith in, and relationship with, God, my naive hope to see combat, and a declaration of my readiness for the sacrifices demanded by the profession of arms; sacrifices the depths of which I could not possibly have understood.

Mom's misgivings aside, she and I had a good time on the road, taking a leisurely drive to Camp Pendleton by way of Las Vegas. We found a cool place for me to live in San Clemente before I took her to the airport and sent her back to Chicago.

Arriving at 3/5 the next morning meant theory was now reality. Checking in to a new unit is like running a gauntlet. It is actually a sizing-up process Marines call "butt sniffing." The arriving Marine must have a very fresh, close haircut and wear his or her green, wool, Class A uniform, called "Alphas," to walk around and visit various staff sections in pursuit of largely meaningless signatures from largely disinterested Marines. Alphas are the institutionally directed choice for check-in because the ribbons and badges worn upon them and the way in which the semi-form-fitting uniform displays your body tells your story, or lack thereof. I had a single ribbon, essentially awarded for breathing while in uniform, perched above the two shooting marksmanship badges I earned at The Basic School. Arriving at a unit with a combat history like Darkhorse's, at a time when most infantry Marines had combat experience in Iraq, Afghanistan, or both, made the lack of experience my uniform belied even more palpable than my baby face. At least Marines could look at me and see that I had a "proper military appearance" and trust that I could shoot relatively straight.

It didn't help that my buddy Robert Kelly had checked in to Darkhorse just a few days ahead of me. Even though Rob was only a few years older than me, I looked up to him as a mentor as much as a friend. He had already endured the 2004 Battle of Fallujah as an enlisted infantryman in First Battalion, Eighth Marines, alongside Darkhorse—the toughest combat the Corps had seen since Vietnam, and he could carry his load and part of yours. Rob was always observing and judging what went on around him and his deeds and conduct in response were flawless. Steven Pressfield wrote, "What does a king do? A king does not abide in his tent." Rob did not abide in his tent; he worked.

Rob had been where I wanted to go, done what I wanted to do, and was already a blooded member of the order to which guys like me, Cam West, Will Donnelly, and the rest of the IOC 4-09 students reporting in to 3/5 were but aspirants. I hoped some of Rob's professional credibility would reflect on me by association.

Checking in also meant meeting the battalion Commanding Officer, Lieutenant Colonel Jason Morris, known as Darkhorse Six, and his Sergeant Major, James Bushway. Fortunately, Lieutenant Colonel Morris met with me and my fellow new lieutenants in one group so my newness was less glaring, camouflaged as it was among that of most of my peers. Lieutenant Colonel Morris was a Marine Corps rock star, destined for general's stars; exactly the kind of officer I had been conditioned to expect by my IOC instructors. Sergeant Major Bushway was five feet, five inches of scars. He was a square human being: square torso, square head, square jaw, and scars. He was . . . terrifying. Again, exactly as I had expected. I would come to admire him tremendously for his perfect balance of loyalty to his commander and unit and his advocacy for the junior enlisted Marines.

Lieutenant Colonel Morris spoke about his command philosophy, the importance of accountability in all things, and the meaning of the battalion's motto, "Get Some!" He also discussed the broad strokes of the battalion's training plan, a conversation that revealed the battalion was deploying to Okinawa, Japan, to serve as the Battalion Landing Team for the Thirty-First Marine Expeditionary Unit instead of combat in Afghanistan. It was my first day in the operating forces, but I was experienced enough to know that a second lieutenant's personal desires were of no consequence. I kept my mouth shut. But I was gutted. I didn't hear a lot of what he said thereafter; I was too deep inside my head.

I had realized so much about myself over the last year, things about who I was and who I wanted to be. I needed my trial. I needed combat to fully realize my perception of myself as a Marine infantryman. I needed it to be validated in the company of people whose opinion mattered deeply to me. After a lifetime in pursuit of achievement for achievement's sake, I wanted to become the fullest version of who I had realized I was. As Lieutenant Colonel Morris finished, I came out of my internal monologue enough to listen as his administration officer walked into the conference room to read off our specific assignments within the battalion.

"Second Lieutenant Schueman, First Platoon, Kilo Company."

I felt electricity shoot down my spine. The disappointment hung like a miasma, but suddenly all the butterflies were gone, replaced with pure anticipation. I wasn't going to combat, but I was going to be in a storied company. This place had a history and I would be a part of it. We rose to attention as Lieutenant Colonel Morris got up and left. As I pushed my chair under the conference table, the Kilo Company executive officer (XO) tapped my shoulder, stuck his hand out, and told me he'd be taking me around.

My company commander (CO) was newly arrived himself. He was in the process of turning over duties with his predecessor. While they talked, I waited outside his office to meet him. In the meantime, the XO introduced me to the Kilo Company gunnery sergeant, Christopher Carlisle. Gunny Carlisle was a big, muscular man with a shaven head. He was in charge of training and logistics support for the company. He shook my hand, welcomed me to the company, and went back to making sure Kilo had what it needed.

The CO came out and welcomed me into his office. We made the usual "where are you from" small talk and then got to business.

"Tom, welcome to Kilo Company. Our call sign is 'Sledgehammer' in honor of E. B. Sledge. If you haven't read *With the Old Breed* I expect all Kilo officers to do so." I was pleased by the call sign.

"Sir, I love that book! I read it on my way out from Quantico."

"We've got some solid training coming as part of the MEU work-up, so you'll get some great opportunities to work your platoon before we deploy. You met the XO and Gunny Carlisle. I know you know Lieutenant West and Lieutenant Donnelly. I expect Kilo to be tight. We're all going to Bridgeport for mountain warfare training and it looks like Korea will happen, too. We will wrap up deployment prep at Twentynine Palms for Mojave Viper. Kilo is going to be ready for war if the North Koreans get froggy."

Lieutenants are expected to be enthusiastic, so I said nothing about my disappointment at 3/5 not getting the call for Afghanistan. Instead I simply replied, "Sounds good, sir. I'm ready to get some."

I told myself I would channel that energy into preparing myself and my Marines for our upcoming seven-month deployment to Japan and the Pacific. We would spend at least three of those seven months aboard a U.S. Navy amphibious ship, stopping to train and party in countries around the Pacific Rim. But the fact that I wasn't headed for combat gnawed at me. The hero's journey demands bravery in service to a greater good, leaving the hero transformed by the experience. I couldn't see myself as a hero unless I was fighting. In 2010, Iraq was believed largely over for Marines. ISIS had yet to make their dramatic rise from the eastern deserts of Syria to march westward toward Baghdad, so Afghanistan was the only place a Marine could find professional and personal validation. The two are wholly intertwined in the Corps. I felt a longing, really a full-blown pang of jealousy every time guys I knew from IOC got deployment orders for Afghanistan. I didn't see myself getting transformed by a bar in Pohang, Korea, at least not into a better version of myself. I wanted the fullness of the hero's journey. I felt it was the only way to prove myself worthy as a man, much less a Marine.

Napoleon said, "A soldier will fight long and hard for a bit of colored cloth." I am embarrassed about it now, but I desperately wanted the extrinsic validation of the yellow, blue, red, and white Combat Action Ribbon, awarded by the Corps as proof that a Marine "engaged the enemy, was under hostile fire, or was physically attacked by the enemy . . . [and] demonstrated satisfactory performance under fire while actively participating in a ground or surface engagement." That ribbon was the measure of a Marine in 2010 and I was without. The absence was like a missing tooth to which my tongue returned again and again. To obsess over a piece of cloth seems absurd, but I was never unaware of the experience I did not have.

When I met my squad leaders, I knew I was blessed with a platoon staffed with experienced noncommissioned officers. First Platoon was a platoon of Marines who joined in a time of war with the expectation and desire that they would fight. They were wild, absolutely savage kids from some of the worst homes in America and they demanded strong leaders. Over time Sergeants Trey Humphrey, Jeffrey Iwatsuru, and Jonathan Decker were the clear leaders in a platoon that was alpha male country. After meeting the Marines, though, it made me wish even more that we were going to "The Show."

Humphrey was my go-to, my main effort squad leader, and a universal fit in any scenario. He could fight, he could kiss babies, whatever. I could give Humphrey a mission and know it would get accomplished to a very high standard. Iwatsuru was always quiet, reserved. He was methodical and careful and by the book. He didn't like to move as fast as I like to move, so there was always a tension between us, but he was an outstanding Marine who executed perfectly. Decker crawled out of a trailer park in the Florida Everglades like an alligator fueled by Monster drinks and microwaved Hot Pockets. He had about the same level of aggression as one. If it had to be destroyed, I would send Decker.

Despite my disappointment with Darkhorse's upcoming deployment, I considered myself blessed to drive the Pacific Coast Highway to work every day. I was going to spend my day exactly where I wanted to be. When I would pass Trestles, the famous surf break, people were usually already paddling out in the dawn light. Pulling through the Cristianitos Gate into the San Mateo area was like landing on an island of infantrymen in the sea of Camp Pendleton. Every day, at 6:30 in the morning, as I drove in I would see Darkhorse Six, Lieutenant Colonel Morris, and Darkhorse Nine, Sergeant Major James Bushway, running toward another area on Camp Pendleton called Camp Talega. Up and moving before the enemy. Weekends were reserved for debauchery with my fellow lieutenants. I wasn't bound for Afghanistan, but I intended to see that First Platoon was the best in 3/5. I was home.

While I was professionally jealous of my friends headed to Afghanistan, we had awesome training planned for our deployment. A patrolling exercise in Camp Pendleton's hills was a great opportunity to immediately get to work with my platoon and my fellow lieutenants and theirs. It also taught me some nontactical lessons.

Officers carried personal two-way radios for training coordination outside of tactical radio networks. I was looking for Cam West and his Third Platoon to link up with them while trying to avoid confirming the stereotypes about second lieutenants and land navigation. Speaking like a normal human is accepted on the personal radios, but since my Marines could hear me, I maintained tactical discipline.

"Sledgehammer Three Actual, this is One Actual, we're heading your direction, can you send me a grid for link-up?"

"Hey, dumbass," Cam responded, "just walk over the hill and you'll see us."

"Roger Three Actual, One Actual copies," I said.

"Hey, shit for brains," Cam responded again. "I'm over here, don't you see us? You need to circle in left."

I responded, "Roger."

Cam met me at the gate of the patrol base. "Hey, you idiot. Here's the gate." One of my squad leaders came to me and said, "You know what, sir? You're acting like we're tactical when you and Lieutenant West are out here bullshitting." He continued: "No wonder why no one's taking this shit seriously; now all the sudden you're all locked on?" It was an embarrassing, but valuable, lesson.

At the Mountain Warfare Training Center in Bridgeport, California, "Sledgehammer" made a name for itself when, unlike the remainder of Darkhorse, we didn't go to ground during a record snowfall that dropped ten feet of snow over seventy-two hours. We looked at the need to overcome the snow as an opportunity and attacked the training schedule rather than delaying it. The decision to go out in the snow and train through adversity was another formative point for me as an officer, a channel for the aggressiveness that IOC had awakened in me, and a decision I would repeat years later when I was in command of a company.

From Bridgeport, Kilo Company went to Operation Freedom Banner in Korea, with truly exceptional training at the Rodriguez live-fire range followed by ninety-six hours of drunken foolishness in Seoul. Right before we headed to Korea, I drew a winning hand when Staff Sergeant Tim Henley arrived to serve as platoon sergeant. It is arguably the most important job in the platoon, as Henley translated my plans and orders into action while providing me wise counsel in generating them. He was a calm, experienced Marine who had served with Dark-horse in Fallujah in 2004 and been wounded there. The story was legend in 3/5. Staff Sergeant Henley thought he'd been hit in the ass with a rock thrown by the kids who followed Marine patrols in Fallujah. He turned to scare them off and then resumed the patrol before someone noticed blood running down his legs and Henley realized the rock was a bullet. He was exactly the kind of man I wanted to serve beside.

As part of ensuring 3/5 was prepared to deploy as part of the MEU, the entire battalion moved to the Marine Corps Air Ground Center at 29 Palms for Enhanced Mojave Viper, a month-long exercise in which the fullness of a Marine Air Ground Task Force is trained and exercised. I had already experienced what it had to offer at IOC and was excited to go back.

Twentynine Palms is unique for its ability to offer Marines the closest replication in training to the wartime ferocity of the Marine Corps, the only training opportunity in which the attack dog that is the Corps can partly slip its chain. Conducting the almost orchestral activities of thousands of Marines moving and shooting and communicating under an umbrella of flame and steel is the essence of Marine Corps operations. We shot thousands of rounds of ammunition, rockets, and mortars. We maneuvered with armored vehicles, helicopters, and tanks under a blanket of integrated live artillery and aviation fires, churning up the dirt and rock of the Mojave Desert. Most importantly, we developed unit cohesion under the closest approximation to combat we could find without being in it.

By December 2009, Darkhorse was three months into battalion-level preparation for our deployment under the command of the Thirty-First Marine Expeditionary Unit (MEU). Training aside, Christmas and the New Year bring the Corps to a halt as Marines and Sailors spend a few precious days of holiday leave with their families. All Marine leave periods begin with briefs designed to limit the liability of commanders and staffs from the fallout from the actions of young men unleashed.

As we gathered for our safety brief prior to releasing Marines for Christmas leave, Lieutenant Colonel Morris stood in a circle of more than one thousand Marines, all impatient for the freedom of the leave period and prepared for the usual tired exhortations about tire pressure and not drinking and driving. Those came as expected. But when Morris announced we would no longer be deploying to Japan, the air became suddenly charged, like the ozone smell before a lightning storm. A thousand Marines collectively leaned in to hear his next words. When he announced that Darkhorse was instead deploying to Sangin District of Helmand Province, Afghanistan, there was a roar so loud I felt as much as heard it.

I could not have wanted anything more for Christmas. I felt a surge of heat through my body, a charge of adrenaline and excitement and, deep down, a tiny, cold pit of fear. I looked over at Rob Kelly, expecting to see the same excitement on his face, and did not find it. I gave him a "Come on, man!" look.

He looked back at me and said with a half smile, "I was ready for a MEU."

Mine was the fire of the unblooded. Rob knew Sangin meant we would bury Marines.

BY 2009, THERE was no question the war in Afghanistan was sliding in the wrong direction. President Barack Obama questioned the estimated $1 trillion cost in blood and treasure to keep Afghanistan moving forward as he contemplated another decade in the country. The depth of the violence in post-invasion Iraq had grabbed America's attention in 2003; now Afghanistan demanded more and more soldiers and Marines to try to stanch the increasing flow of blood; more and more bodies to feed into Ares's gaping maw.

The commander of the International Security Assistance Force (ISAF) in Afghanistan, General Stanley McChrystal, completed his Initial Assessment and released it to Secretary of Defense Robert Gates on August 30, 2009. The upshot was a request for a "surge" of 40,000 more troops to implement the kind of robust counterinsurgency he saw as critical to a successful conclusion in Afghanistan. He and U.S. Central Command commander General David Petraeus were publicly vocal about the need for the surge. That was a break in the norms expected of the military, particularly general officers, and it put them at odds with the White House.

Then–Vice President Joe Biden wanted a "counterterrorism-plus" strategy: a few thousand special operations troops backed up by aviation and unmanned aircraft to keep an eye on things and play Whac-A-Mole with terrorists. General McChrystal publicly discounted that plan in Paris on October 1, 2009, further driving a wedge between the uniformed and civilian contingents and angering President Obama, who felt the military was trying to paint him into a corner.

Ultimately, President Obama decided to send 30,000 additional reinforcements to go hard against Al Qaeda, as Vice President Biden wanted, and further develop the Afghan National Army's capabilities to defend against the internal threat posed by the Taliban by advising and assisting them in battle, as McChrystal and Petraeus had lobbied. That brought total troop numbers in the country to 100,000. President Obama also announced that the American military drawdown would begin in July 2011, because he didn't think the American people would support a longer presence. Critics like Senators John McCain and Lindsey Graham argued that setting a date for withdrawal gave the Taliban a date against which they just needed to endure. It was a fair critique. Endurance is a specialty of the Afghan people.

Twenty-one thousand of the surge troops went to Kandahar and Helmand Provinces. The Taliban controlled most of Helmand by early 2009 and the provincial capital of Lashkar Gah was under pressure. Marine Corps Commandant James T. Conway was the leader of a force without a fight, anathema to the Corps, and he lobbied for the job of solving that problem. As Captain Willard said in the movie *Apocalypse*

Now, lying drunk in a Saigon hotel room, "I wanted a mission, and for my sins, they gave me one." Helmand belonged to the Marines. In Kabul and the Pentagon, the powers that be called it "Marineistan."

Marine Corps doctrine specifies the employment of Marine-pure task forces built from units with habitual working relationships. The rosters are built specific to the requirements of each deployment. From our perspective in 3/5, Marine ownership of Helmand offered a chance to fight in the way we'd practiced at Enhanced Mojave Viper, as part of a Marine Air-Ground Task Force, using tactics, techniques, and procedures we'd practiced until they were muscle memory.

The Taliban leadership responded by ordering their foot soldiers to reduce casualties by avoiding direct firefights with U.S. forces in favor of the use of improvised explosive devices (IED) and small ambushes to inflict casualties with minimal risk to themselves. Beleaguered British forces, experienced in counterinsurgency in Northern Ireland and southern Iraq, had been in Helmand for three bloody years and found a culminating point there. Still, they were willing to help the incoming Marines acclimate to the area of operations.

I watched all of this transpire with something approaching professional interest but more akin to the anticipation of a college athlete hoping to make the pros. I was aware of only the most basic details of the grand strategies, but I wanted to go to war as part of my journey. The larger reasons for it, and the prospects for success, were really not my focus. The waste I now know as endemic to war was not my focus. I was focused on rifle marksmanship, the movement of my platoon from one point to another to accomplish an assigned task, and accountability for the off-duty actions of men in their early twenties, jacked up on testosterone and invincibility. All Marine infantrymen have "fear of missing out." I just hoped there was still something left for me to do.

WITH A COMBAT deployment looming, my personal life remained a mess, if only because I made it that way. I was still suffering from my own weaknesses. Andrea and I had remained in a date-and-breakup cycle still largely driven by my own shitty behavior—my default response to commitment. Terrified of marriage, I was all over the place. But on the cusp of shipping out in September 2010, I decided to take her ring shopping three days before deploying.

We spent hours at a jewelry store looking at rings, but in the end, I failed. My cowardice was as much couched in my fear of committing to something real and good as any intent to protect her. A better man would have dropped to a knee and asked her to marry him in the jewelry store. If there was any nobility at all, it was in the notion that after the way I behaved during ten years of dating, and if I was unwilling to get engaged to her now, it was unfair to expect her to wait on me as a boyfriend while I was in a place that could only cause her fear and worry.

IOC had focused all of us on the idea of conducting counterinsurgency operations. At 29 Palms, Enhanced Mojave Viper was centered on the mission statement coming out of the ISAF Headquarters in Kabul to "shape, clear, hold, build" in Sangin. Shape the environment for operations against the Taliban. Clear the Taliban from the Sangin Valley. Hold the area against their resurgence. Build a viable representative government as an extension of the central government in Kabul. In other words, much of our training focused on culture and winning hearts and minds; we thought we might fight, but we really thought we were going to be kissing babies. Then, two weeks before we deployed, Captain Ryan Cohen came to talk to us about Sangin.

Cohen was the company commander for Kilo, 3/7. He had been medevaced from Sangin after being wounded a second time. He was a modern-day Spartan warrior and he was blunt: "We're fighting every day. There are IEDs everywhere. Marines are getting blown up regularly." Everyone had been acting as if we were going to Afghanistan to dig wells. Cohen's talk was the first indication that we might be facing a different, far deadlier beast.

The next day I drove Andrea from San Clemente to Los Angeles International Airport, where she would fly home. As we pulled into the departures lane, I panicked. Afghanistan, and what Cohen had told us, was a useful cover.

"Listen, I don't know what's going to happen on this deployment. You should find someone else."

"What?"

"Find a boyfriend while I'm gone. We're not getting married. This is over."

There was no time to argue. I set it up that way. She looked at me, furious and deeply hurt. I saw the resolve harden in her eyes as she dumped a Coke in my lap. Then she was slamming the car door.

"You're a jerk, Tom! I cannot even believe you. You've ruined my *life*!"

I couldn't even look her in the eye as I got out and yanked her bag out of the backseat for her. I watched her walk away, and half-hoped she would turn around. She didn't. I drove straight back to San Clemente.

Cam was his usual self, helping me handle a major life moment by making fun of me and making me laugh at my own idiocy. Cam is a John Wayne figure, if John Wayne were the John Wayne of the movies. He's six foot three and everything he does is big and loud. He led Marines by sheer force of personality, and they loved him for it. If I'd followed through on getting engaged, he would have celebrated with vigor. As it was, he told me I had done the right thing and reminded me that I was too stupid and immature to get married anyway. That night he knew there was time to wallow in misery over what I had done. Instead he led me out to get very, very drunk.

The next day, my phone rang at 9 A.M. The sound was like a spike driven directly into the center of my hangover. It was Rob Kelly.

"Hey, come down to Del Mar," he said. "We got a cabin down here, my dad is grilling, come over for lunch."

With about twelve hours left in America, Cam and I headed down to Del Mar on a shockingly bright, sunny Southern California day, with the type of hangover you have when you dump your almost fiancée the day before you go to war. It only got worse. I tried to nurse a beer but

Rob's dad, General John Kelly, was feeding me Pabst Blue Ribbon. The only answer in that case is "Thank you, sir, may I have another?"

I initially felt physically exhausted from the previous night and emotionally distant from the moment. But it was the kind of Southern California day people write songs about and a beautiful day spent with a beautiful family fully oriented on their son pulled me into their orbit. Seeing the love the Kelly family had for Rob, and the way his decency was imputed to Cam and me, made me feel accepted into their family. Rob was making us all laugh with his dry, snappy humor. The day revolved around Rob as the son deploying to war, but they loved all of us as Marines, as members of the family business. They also so clearly loved Rob for himself, for what he gave and the essential goodness with which he represented them all, something each of them did in return. They were a family wholly contained within themselves. They were enough for one another, and all each of them needed. There was no yelling, no one was going to throw anything, no one was going to lose control of themselves and ruin the moment. They were confident enough in themselves as individuals that they had surplus love to give everyone else. The fullness of that kind of love was something I had rarely experienced.

With the sun setting on the Pacific, I sat with Cam and Rob. Afghanistan was a weight on our shoulders. Rob stood. He walked from the porch down to the beach. Cam and I followed. I thought about war and felt nervous anticipation. I thought about Andrea and felt guilt. I figured I had gotten some of what I wanted and some of what I deserved. As we watched a last sunset, I'm sure there was some talk of what was coming. But in my mind, there are no words in that moment. We just stood and watched as the sun slipped under the water, leaving the sky ablaze in red and yellow and orange, then turned and walked quietly back to the party.

A few hours later, I was on my way to Afghanistan.

7

Welcome to First Platoon

ZAK

I WOKE UP at FOB Inkerman for the first time on October 14, 2010. I made my preparations for prayer. Kamar and Adan joined me. There was no running water, so we cleaned ourselves with a bottle of water and then offered praise to Allah. I was hungry, so I ate one of the halal MREs in a box I found under a bunk. On the outside, it said it was chicken and lentils. I thought about my mother making fresh naan and green tea.

I did not know what the day would bring so I reached into my bag and put on the camouflage uniform the Marines gave me at Camp Leatherneck, the same one they wore. I put on the boots they gave me, too. They were new and stiff. I wondered how far I would have to walk in them. When they gave the boots and uniform to me at Camp Leatherneck, they told me I would dress like the Marines on patrol so I did not stand out. Before Leatherneck, I did not know that we would be working with Marines. I knew some Marines five years before at Topchi in Asadabad, but mostly the U.S. Army was in Kunar. I really didn't know the difference. I learned in Sangin.

The night before, the staff sergeant that brought me to FOB Inkerman said that he would be back to take me to meet my platoon that morning. I looked outside, but there was nothing to see of Sangin. The FOB was surrounded by mud walls and the same HESCO boxes I had seen at Topchi and FOB Blessing. The ground was the same ankle-deep dust as at Leatherneck. I did not know anyone else at FOB Inkerman, so I stayed in my room and tried to arrange my things and prepare for whatever was coming. I did not know the staff sergeant's name or where he stayed. I did not want to miss him if he came to get me.

Around 9 A.M., the staff sergeant came back.

"Mornin', Zak. Grab your trash and let's go. There's a patrol going out and they need you." Just like that.

I was suddenly nervous, and my heart beat faster. I thought, "It is time to do what you said you were going to do." He showed me the things I would need to carry in addition to my one bag of clothes, just my helmet, body armor, some water, and a little food, then I followed him to where First Platoon lived.

It wasn't much different than where I had slept, buildings made of mud bricks, rooms with no windows. There were Marines outside, standing under camouflage netting while they put on their equipment for the patrol. They were all my age. I thought they would be older. They carried rifles and pistols and machine guns. Some of them carried what I would later learn were metal detectors for finding IEDs. As they dressed, they looked each other over, helping with the equipment they carried, making adjustments. The staff sergeant walked me past all of the Marines to a tall, thin man. He introduced me to him. "Zak, this is Lieutenant Schueman. He's your boss. You go where he goes and do what he tells you." He turned to Lieutenant Schueman: "Sir, this is Zak. He is your translator. I've got his gear set up to patrol, he's good to go."

I felt like I knew the Lieutenant as soon as I saw him. Looking at his face, I said in my head, "He's a good man." Lieutenant Schueman stuck out his hand to shake mine, firm like the Americans do. He had a big smile. "Zak! Welcome to First Platoon. We are going to patrol from here at Inkerman to Patrol Base Fires. You ready to get some?"

I didn't really know what he meant, but I thought there must only be one answer, so I said, "Yes, sir!" I felt happy and comfortable because he seemed like a nice person and was respectful to me. He turned to the platoon.

"Hey! First Platoon! Listen up! This is Zak. He's our interpreter and is going to help us get some work done. He's one of us now. Staff Sergeant Henley, get his information on the kill sheet. Any questions?" No one had any questions. Some of the Marines came and shook my hand. They said things like, "What up, Zak?" or "You ready to whoop it on?" I was nervous because I was still working on my English, but after that, I felt accepted by the platoon. Like them, I had a job to do, and I was proud to be there doing it. I had been thinking about the fighting in Sangin. Now I was about to go out in it.

Except for the IED that hit my truck, when I was shot at in Nangalam, it was always from far up on the mountainsides. Helmand was flat and dusty except for the agricultural areas near the river where corn grew high and the Taliban could shoot and run before we even knew they were there. Even worse, they could set IEDs everywhere. Eight Marines from 3/5 were killed in the week before I got to FOB Inkerman. I did not want to die or be hurt but someday I hoped to have children and I wanted for them the freedoms I did not have when I was a child. I could not allow the Americans to pay all of the cost of that. I stood next to Lieutenant Schueman and waited for him to tell me where we were going.

WE LEFT FOB Inkerman for my first combat patrol at 12:30 in the afternoon. The sun was high overhead and the temperature was in the eighties. Lieutenant Schueman and I were going to patrol with Third Squad of First Platoon. Sergeant Jonathan Decker was the Third Squad leader. He came to speak to the Lieutenant about the patrol before we left. As I put on my backpack, Decker turned to me. He pointed his finger at me and said, "You watch where the Lieutenant puts his feet. Do not step anywhere he did not already step! Do you understand?" I said I did and followed them out through wire and HESCO barriers. I tried not to show the Marines that I was scared, but I think they knew.

I kept my eyes on Lieutenant Schueman's feet. He stayed on hard-packed ground wherever he could. The squad moved separately. One part moved forward and set up their guns so they could watch the second part move forward and protect them from the Taliban while they moved. That was how Marines often moved in Sangin, but I did not know anything about that at the time. I heard Sergeant Decker say on the radio, "Hey, Gonzales, start moving up." I was not even really aware of what was happening, I was just focused on Lieutenant Schueman's feet. I was watching him so closely I did not understand when I heard a sound far away, "THOOMP . . . THOOMP! THOOMP!" then "CRUMP . . . CRUMP! CRUMP!" as puffs of smoke and dust rose in a field to our left.

I really did not understand what I was seeing. All the Marines were lying down on the ground, shooting back. I was frozen. I was standing up, just looking around. I was wondering who had just fired at us. I did not know what to do. I had never been in this situation and I had no training for fighting. This was different from Nangalam. This was not shooting from a mountainside that might hit someone. Here in Sangin, the Taliban were trying to kill me and the Marines I was with. There were about four Taliban in a tree line three hundred meters away from us. They were shooting 30 mm grenade launchers at us, but they were not landing close to us. Then they began shooting with AK-47 rifles and PKM machine guns. Every Afghan knows the sound of those weapons.

One of our corpsmen, Doc Doug Long, was yelling at me, "ZAK! LAY DOWN! LAY DOWN!" But it was like I was deaf and frozen in place. Time was moving slowly. Doc Long reached up and grabbed the handle on the back of my body armor in his hand. He slammed me to the ground. He was on top of me, screaming in my ear, "STAY DOWN!" I pushed my face into the dirt. "Okay! I am laying down!" I was scared but I was thinking, "This is a war. You know you are going to fight. Whatever happens to the Marines, it's going to be the same for you, too." I still feel like that.

Doc got off me. He had his head up, looking for injured Marines. Everyone was yelling. I heard Sergeant Decker calling on the radio, asking for mortars at FOB Inkerman to fire at the Taliban. From Inkerman I heard a sound like the Taliban grenade launchers, but louder. The sound of the Marines firing their guns at the Taliban in the tree line made it hard to hear but the mortars were loud. I saw small bursts of fire and a sound like doors slamming in an empty house as the mortars exploded near the tree line. Marines were cheering as the explosions ripped leaves from the trees.

As the Taliban stopped shooting at us, Marines started standing up again and the part of the squad separated from us started moving to us. The Taliban began firing again from our front and left side. Lieutenant Schueman called out, "Get some rockets on that tree line!" Two Marines ran forward with a rocket launcher. One knelt down. The other loaded a rocket and slapped the kneeling Marine on his helmet. The kneeling Marine shouted, "ROCKET!" and pulled the trigger, but nothing came out. The other Marine reloaded and slapped the Marine with the launcher on the head again. "ROCKET!" Nothing. Sergeant Decker looked at them and shouted, "MOTHERFUCKER!" He looked angry as he grabbed his radio handset again. He spoke into the black handset and called Inkerman for more mortars, which soon started falling. The Taliban did not shoot again after that.

We started moving again. I thought maybe we would go back to FOB Inkerman and rest after the firefight, but Lieutenant Schueman just looked at me and spit some Copenhagen as he smiled and said, "Let's go. We got work to do."

We spread out the patrol and started moving as separate teams again to cover each other. I never got to speak to an Afghan that day. I just got shot at by them. We walked and walked, but at the end we were back at FOB Inkerman, where we started. The platoon seemed satisfied with my performance in the fight. We had been shot at, but everyone was okay. Another platoon in Kilo Company was not so lucky. A Marine named Lance Corporal Alec Catherwood was killed by an IED. Like me, it was his first patrol. Three more Marines were wounded. Two of them lost their legs. I wondered if every day would be like that.

NOT EVERY DAY was like October 14. But many were. Many days seemed like the same day. The only difference was whether we killed the Taliban or they killed us. There was another interpreter at FOB Inkerman. I don't remember his name, but he came from America. He worked inside the FOB. My team needed me because I was the only interpreter for First Platoon. I was the interpreter for outside the FOB. I walked so many patrols. My job was to talk to the people of Sangin, to tell them what Lieutenant Schueman or a squad leader said and tell the Marines what the people said.

I was enthusiastic to serve my country and work together with the U.S. and international forces, but I also felt I worked for the people in Sangin, who had no way to have a good life. They were starving. They had so much hardship. The Marine forces were there to secure the province and try to make things better for them. I was, too.

The people in Sangin had different accents than I was used to in Kunar. It is the same in America: different places, different accents. But we were all Muslims. We were from the same country. We were all Pashtuns. The Koran says, "If anyone relieves a Muslim believer from one of the hardships of this worldly life, Allah will relieve him of one of the hardships of the Day of Resurrection." I wanted to help them during our missions. I wanted to keep them from getting hurt or their homes damaged. And we did. We helped people a lot, especially by clearing the area around the villages of IEDs. The Taliban would place the IEDs in trails and roads or anywhere they thought we would walk. First Platoon would find the bombs and destroy them or deactivate them whenever we could. Still, there were many people in Sangin, especially children, who were missing legs. They had little chance for a good life in Afghanistan after that.

Some of the people in Helmand called me infidel like the Taliban did. I did not listen to them. I knew I was true to my religion. I did my ablutions before the patrols, so that I could pray during the patrol. When it was my prayer time, Lieutenant Schueman would tell all the Marines to stop the patrol and find a secure place for us while I prayed. Sometimes the Taliban shot at us while I was praying. So, who is the infidel?

Highway 611 was the main road for people in Sangin to get to the district center. It ran through the center of First Platoon's area. The Taliban placed IEDs in the road, especially the unpaved part of it. We tried to secure miles of the road by taking over old buildings as observation posts along the road so we could watch and patrol to keep the Taliban from placing bombs in the road during the night. Lieutenant Schueman and I would talk to the owners of compounds and ask them if we could stay there in return for money. He was respectful to the people and that made me care for him even more.

I was always with Lieutenant Schueman. When we met with the village elders, I sat next to him and explained what each person was saying. I also helped him understand what they meant. They are not always the same thing in Afghanistan. Sometimes we were soft, talking about schools and roads. Sometimes we were hard. I had to reflect Lieutenant Schueman's mood and intent in communication with people who we knew were Taliban. Sometimes it was uncomfortable for me, but First Platoon was my team and I supported them.

"Zak, tell this guy things are going to change around here." I translated that for the elder. He spoke back to me, his long white beard shaking as he spoke.

"Sir, he says, 'That's what the British said, that's what the last Marine unit said. The Taliban owns this village, you will never be welcome here, you'll never be here.'"

"You tell him I'm going to kill every single Taliban until this village is no longer Taliban." The old man smiled. He had less teeth than I have fingers. He replied and I translated.

"Lieutenant, the man says, 'You can kill all the Taliban, but then you have to also kill all their wives because their wives will also fight you.'"

The Lieutenant looked at him and said, "You're going to have a lot of dead women, too."

At FOB Inkerman, there was a contingent of the Afghan National Army (ANA). The Marines were supposed to slowly hand operations over to them. They did not always want to work, so I would make them support the Marines by saying, "These Americans are doing things for your country you should be doing." Sometimes Lieutenant Schueman would eat with them. I always went with him to translate and have real Afghan food instead of MREs. The ANA soldiers brought me Afghan tobacco. The Marines all used Copenhagen or Grizzly snuff. They called it "chew." I shared my Afghan chew with the Lieutenant and the other Marines. It was much stronger than theirs. The Marines said it was too strong for them.

During patrols, I walked just behind the Lieutenant, far enough away that an IED would only injure or kill one of us. In the field he helped me learn everything I needed to stay comfortable. He gave me extra clothes because I did not have much and my clothes ripped in the field. When it got cold, he gave me a jacket. He gave me a Marine T-shirt I kept for many years.

A lot of the time in Sangin was sadness, but there was a lot of fun, too. Sometimes they came together. Even the fighting was fun sometimes. When the Marines killed a Talib on a motorcycle five hundred meters outside our patrol base, I ran outside and got the motorcycle and rode it back to the compound. The Marines made a racetrack inside the patrol base and we had contests to see who was the fastest. Later Lieutenant Schueman was upset for me. He did not know that the Marines asked me to get the motorcycle or that I was outside. He told me that the balloons with cameras in the sky were watching me ride it back and thought I was Taliban. They were going to shoot me with a helicopter. Sergeant Decker heard them talking about it to Lieutenant Schueman and yelled, "It's Zak!" Lieutenant Schueman stopped Kilo Company from shooting me. It only took one mistake to die in Sangin.

Before I got to Sangin, I was worried about what fighting would be like. It was hard the first time. But I got used to it. While we were patrolling, I listened to the Taliban on the radio. They knew I was listening. I could hear them following us as we patrolled. They passed us off to other Taliban as we moved farther. The Taliban talked about IEDs in code. Their codes were simple. They called IEDs melons or pumpkins. They would say, "Is the pumpkin near the road?"

"Yes! The plate is in the road near the pumpkin, praise be to God."

"Okay! I see the Marines, they are coming to it."

Listening to them let me warn the Marines. I could tell my team, "Be careful, there's an IED ahead" or "They're sitting in an ambush near the bridge." But sometimes the Taliban knew we were listening and they called to me, "God willing, we will catch you, infidel! We know your name, Zainullah Zaki! We know you are from Kunar! We condemn you for helping Americans, infidel! We will cut off your head after we torture you!"

I had only been in Sangin for four weeks. But there were locals who worked on FOB Inkerman and Afghan soldiers who lived on FOB Inkerman who talked to the locals. Through that word of mouth the Taliban knew my name and where I was from. Interpreters were not supposed to talk to the Taliban but I could not help myself. I told them, "Fuck you! Bring it! We will take care of you no problem! We will kill all of you!"

Usually they did not know we were listening. One time, as we approached a village through a cornfield that had been harvested, I could hear the Taliban on the radio. They were describing our movement and planning an attack on us. I called to Lieutenant Schueman, "Sir, the Taliban is setting up an ambush. I know where they are from how they are describing us." He looked surprised.

I grabbed his map and showed him: "They're talking about it here, they're talking about it there." I pointed to some houses on the map and matched them to the houses we could see. "Sir, for them to be talking about movement and everything, they have to be in one of these buildings over here. They see us coming in. They are talking about it. We need to go. They are watching us right now! We need to get there before they start this ambush!"

Lieutenant Schueman had the whole team to worry about. They moved slowly because they had to sweep and look for IEDs in the ground.

"Zak," he said, "I'm moving as fast as the engineer can sweep for IEDs! We can't walk over dirt that hasn't been swept. You know there are pressure plates in the ground between us and them! That's why they want us to cross it and I'm not getting anyone blown up!"

I knew that if we kept moving slowly the Taliban would start firing at us. I could tell from what the Taliban was saying on the radio just which house he was in. If I could not make the Lieutenant understand that an attack was coming, I had to protect my team.

I sprinted toward the house, passing Marines on the trail running between the cornrows. They were shouting at me to stop, every Marine passed joining in yelling at me to come back. Green cornstalks rippled behind me, a signal of my progress toward the houses where the Taliban hid. I wished I had a weapon, but interpreters were not allowed to carry them.

I knew where the guy would be. It was a brown house made of mud brick, like every other, but set apart and offering the Taliban a good view of the field where the Marines slowly worked their way over the ground. I broke out of the corn and ran across the tilled earth of an open field. I hoped there was no IED, but if today was my day, Allah willed it.

When I got to the house, I could hear the words of the person inside and knew the house held a Talib talking to others on a radio. I tore open the door and saw the bearded man I expected. He had a radio in his hand, held to his mouth, and looked shocked to see me. He was small and skinny, like most of the people in Sangin. I did not say anything. I just charged him and hit him like an American football player.

With my body armor and helmet on, I outweighed him. He collapsed under me and his radio went skittering across the floor. He did not fight hard. I only had to hit him once before I pulled him up and put the radio in my pocket. I grabbed him by the neck and pulled his arm behind his back as the Marines taught me and walked him back across the field toward the Marines. They had started moving faster when I ran by them and were coming across the field. I told the Talib he was lucky I got to him first. A Marine sniper would have killed him immediately if he had seen him talking on the radio through a rifle scope.

Later Lieutenant Schueman told me that I was crazy to run through the field without the minesweeper going ahead of me. I knew that. But everywhere in Sangin was a minefield. I knew where the Taliban was that was talking about us on the radio. I knew the Marines could not get there in time to stop him, so I just ran to the building and got him myself. My team could not risk doing that, but I could not risk the Taliban getting to hurt my team. We had lost too many already.

We were all young men of the same age. I was twenty like most of the Marines. Lieutenant Schueman and Staff Sergeant Henley were a few years older than us. We laughed and talked about things young men always talk about: food, women, sports, and the world we saw around us. They were curious about my life growing up in Afghanistan. Kunar was not Sangin, but they were separated from Sangin by wire and walls, so I was Afghanistan to them. I tried to answer their questions about the country to which they had come; where some of them would leave parts of themselves and some of them would lose everything.

Marines argue constantly. They compete about everything. They had contests to see who could throw the heaviest rock, throw a rock the farthest, throw a rock the most accurately. They captured mice and raced them on a track. One Marine made a trap out of a box and stick with a string. He baited it with cheese and crackers and then sat for hours waiting on a mouse to come. While he sat he told me how much he loved fishing near his home. I told him about the fish in the Kunar River and how sweet they taste.

They told me about America and all the different places and different ways they lived. Like me, most had recently graduated from high school. They liked to talk about cars and motorcycles. I had never had a car of my own, but many of them showed me pictures of theirs. We all complained about the food at FOB Inkerman. I did not like MREs and neither did they. We all missed the food our mothers made. No matter where we came from, now we lived within feet of one another, so we all lived the same life. We became friends and brothers.

For a young man with no wife or kids, fighting could be fun if no one on our side got hurt. Young men like excitement, and combat *was* exciting. But it was still horrible. After only six weeks, I had seen so many bad things with Lieutenant Schueman. Many civilians got hurt because the Taliban did not care. Many Marines got hurt trying to help the people. Because he was our leader, Lieutenant Schueman could not talk about fear or sadness. He treated me like a Marine, but I was not a Marine, so we could be friends. We could talk about things. Many of his friends were hurt or killed. The longer the violence continued, the more he needed to talk. So did I.

Away from the Marines, we spoke freely, like I was talking to my older brother. I shared the same dangers, standing next to him. He could explain and reveal how he was feeling and he could trust me because I was feeling the same things. We both wondered if we would survive Sangin. We both wondered if we would see our families again. We both wanted to know whether what we were doing had value to the world.

Of course, we talked about our families and how we missed them. I wanted so much for myself, my family, and for all Afghans. Lieutenant Schueman knew a lot about the American Constitution and he told me about it. I was shocked when I learned that many of the Marines had not ever read it. The Lieutenant and I discussed how the ideas within it could work as part of Afghan culture. I wanted freedoms for myself and the people of Sangin. But many people in Afghanistan had no concept of those basic freedoms that I wanted. They were not woven into the fabric of life in the villages.

Afghans value their families, their tribes, and their villages. A lot of people said, "The president of Afghanistan is really the mayor of Kabul." Much of what happened in Kabul did not get to the villages, and American ideas like "one person, one vote" are easily crushed by local warlords or people with enough money to buy votes. Bribes make normal commerce hard, as Americans understand it. Lieutenant Schueman helped me understand American politics. That mattered because American politics are the world's politics.

We talked about religion, too. Lieutenant Schueman explained what he believed about God and the importance of those beliefs to him. He explained that in his faith, he was unafraid to die. He knew he would go to Heaven because his sins were forgiven by Christ. I talked to him about Islam and how important it is in our daily life, how I could also spend eternity in Paradise by living according to the word of Allah and asking for his forgiveness for my weakness. I enjoyed finding our similarities.

Sometimes we just made fun of each other or the people around us. When so much around you is so serious, often you need to laugh more than anything else.

8

Sledgehammer

TOM

FROM THE BATTALION headquarters at Forward Operating Base (FOB) Jackson, 3/5 was responsible for almost thirty-seven square miles of Sangin District and almost sixty thousand people living throughout. Kilo Company was headquartered at FOB Inkerman, four miles northeast of Jackson, and responsible for a smaller piece of Sangin stretching almost another four miles northeast to Outpost Transformer. A mile and a half north of FOB Inkerman and the same distance west of Transformer lay Patrol Base (PB) Fires, from which First Platoon was assigned to operate. I was in Afghanistan a week and at PB Fires for twenty-four hours when I received my first mission on October 9, 2010. I had everything I thought I wanted. Now it was time to find out the truth.

I left PB Fires with a mix of ninety-six Marines and Afghan National Army (ANA) soldiers to look at three areas the 3/5 headquarters wanted to know more about. In addition to First Platoon, I had a host of attachments: four Marines from 3/7 to give us their insights into the area and their experiences within it; an entire squad of engineers and two Explosive Ordnance Disposal Marines to contend with IEDs, the primary threat in Sangin; a section of six snipers; the Embedded Training Team (ETT) detailed to work with the ANA; three interpreters to help us communicate; and a Joint Terminal Attack Controller (JTAC) to get us air support if any of the Taliban, who typically operated in groups of three to four men, wanted to challenge a group of ninety-six.

Sangin was primarily dirt and blowing dust, but irrigation canals stretching out from the banks of the Helmand River created verdant fields of corn and the pink and red splashes of opium poppy. On trails, roads, anywhere we would be forced to pass, the Taliban hid IEDs by the hundreds. They buried yellow plastic jugs containing pounds of home-made explosive triggered by pressure plates that completed a circuit when someone stepped on them. IEDs were indiscriminate weapons, killing Marines and Afghan civilians alike. Their density demanded we move single file, with about thirty feet between each person, to keep an IED from getting more than two or three of us at once. PB Fires was at the center of it all.

Walking single file is the best way to mitigate an IED threat, but it is a horrible way to encounter your first ambush. I was at the front of the formation, moving among the jade leaves and whispers of wind through a cornfield, stalks towering over us all and concealing our movement, when machine gun rounds from a nearby building began ripping through the corn. I dove behind a brush pile that concealed me from view but would not stop a bullet, much less hundreds of them. The size of our group and the dispersed formation meant our element was in a firefight before the remainder of the patrol completely exited PB Fires. I was going to die before I got the second half of my unit out of the base.

I looked to my left at one of my Marines, Sergeant Joseph Nikirk, and shouted, "I'm going to stand up and put suppressive fire through that window. You fire a grenade through it!" I stood and fired at the window as quickly as I could while still retaining any kind of accuracy. Nikirk followed with a grenade that silenced the machine gun. We both dropped back into the corn, where the 3/7 lieutenant acting as our tour guide turned to me and said sarcastically, "Lieutenant Schueman, you got your Combat Action Ribbon, now make a decision!"

I was angry. He was a fellow lieutenant, not an IOC instructor. I was the new guy so I stayed quiet and took it. But I did think to myself,

"Make a decision? I did make a decision! My decision was to not die in the impact zone of a machine gun and to kill the guy shooting at us." I thought I wanted combat, but finding myself in a complex ambush with half the platoon still in the patrol base seemed like a bad way to start. Suddenly the Combat Action Ribbon seemed less important.

Less than an hour later, I knelt at the edge of a cornfield as the ANA, their American advisors, and the three interpreters spoke to a local elder to tell him we would need to use his compound for twenty-four hours. Like most in Helmand, it was a mini-fortress made of thick mud walls six to ten feet high with a sheet steel gate. Goats and children would almost certainly roam within while women in burqas stayed out of sight inside homes made of the same mud brick as the walls.

As I waited for an update from the compound, a single shot rang out, followed by Sergeant Decker coming to me at a run. "Sir, Teague shot someone." I took a breath.

"Okay, did the guy have a gun?"

Decker nodded his head. "Yes, sir!"

"Is he dead?"

"Very."

"All right. Well, tell Teague he did a good job." That's when the reality of Sangin hit me. We could kill people here. It was a notion that would soon become prosaic, but in the moment, it was striking.

The ETT Marines returned from talking to our prospective landlords. "We can't stay here," they said.

I was taken aback. "What do you mean we *can't*? We *are* staying here."

The ETT leader, Lieutenant Chip Broun, said, "The dude that owns the place said the whole village is staying with him and it would be suicide for them to try and go anywhere else. He appealed to the ANA and some cultural stuff went down. I don't know, man. The ANA are saying we shouldn't evict them."

I had practiced this kind of scenario at 29 Palms. But there, diplomacy was academic. "Go back and tell him that if we can't stay at his house, then he's going to take us somewhere that we can stay." Our emissaries walked back to the compound and soon returned with a guide.

We ended up at another compound a few hundred meters away. Corn bobbed in the breeze, so close to the short outer walls that it rubbed against them. It violated everything I had been taught about a defensive position. The proximity of the corn concealed any enemy approach and inhibited the effectiveness of our weapons. I didn't like it. I liked it even less when, as Marines searched the compound, I heard someone scream, "*Grenade!*" I dove for cover as it exploded.

I was not even certain how all ninety-six of my people were arrayed in the corn. Now someone had slipped through them and gotten close enough to pitch a grenade over the wall at us. I looked at the ETT leader for a third time. My mask of diplomacy was slipping. "Chip, go tell that asshole he has one more chance to find us better digs for the night or we are staying at his place. And next time I won't be asking."

As I waited for them to return, feeling like someone on a bizarre version of one of those "find a new house" shows where people look at three places and choose one while suspenseful music plays and the hosts stare expectantly, I heard three loud thumps. I keyed my radio handset and shouted across the airwaves, "Is someone firing mortars?"

The voice of one of our mortarmen came back. He sounded nervous. "Uhhh, yes, sir."

I had no idea what they were firing at, or why, and they had no answer for me. I did know that our company commander was the only one who had the authority to approve a 60 mm mortar mission.

Before I could even think about how I was going to explain that, I heard the muted crack of one of our snipers firing his suppressed rifle and killing a Taliban fighter, the first of many sniper kills over the deployment. I was about to key the radio and asked who the snipers were engaging when the advisor team lead came back with a local resident. I clarified my understanding of the sniper's actions and then we followed the local guide to a third location.

The third location was suitable. There was no crown molding or open-concept floor plan, and the bloated dead goat in the middle of the courtyard caused some initial hesitation, but the walls towered above us and no one was home. Perfect.

As Staff Sergeant Henley put Marines in position to provide security overnight, Third Squad departed on a patrol to examine the impact area of an artillery strike. The battle damage assessment (BDA) of the area was the first of the three that 3/5 wanted us to check out. I went with Sergeant Humphrey and First Squad to answer another question about a compound believed to be a Taliban meeting site. Briefings we received, and my own inexperience, left me convinced the Taliban were ten feet tall and that fighting them in a building meant certain death, so we watched the compound from a tree line and returned to our own after an hour. My overcautious indecision still haunts me. I am confident the compound was a weapons cache that supported attacks on us for months to come. Searching it may have saved lives.

The last area we were to check was a Taliban commander's house in an area never patrolled by U.S. forces. I waited until nightfall to go. There was no moon and no electricity. The entire district disappeared into the night. The corn was so thick and the night so black you could find yourself alone—lost in an instant. Dogs howled as we slipped in and out of chest-deep canals to avoid IEDs, the water warmer than the air. My fear of a break in contact, along with my wet cammies, left me shaking. We reached the commander's house around midnight. We weren't allowed to search houses at night, so we observed briefly before slipping back into the canals and stumbling our way back through the corn.

Back at our temporary lodging, I radioed Kilo Company Headquarters at FOB Inkerman at 2:00 A.M., "Sledgehammer Main, this is One Actual. All Marines present. We are going firm for the night. I am low on batteries so I'm turning my radio off until 0400, at which time we will break down our position, and return to base." The watch officer acknowledged my transmission.

I checked on the Marine standing post outside the steel gate, then peeled off my soaking wet uniform before I lay down next to the radio, my water-wrinkled extremities and chattering teeth rolled in a thin nylon blanket. I was soon unconscious.

"Tom! Tom!" Another lieutenant accompanying the patrol was shaking my shoulder. I snapped awake. "What?!"

"I didn't bring anything to sleep in." Lord give me strength.

"Okay, dude, you can share my poncho liner." I stood. "Dude, you still have your soaking wet cammies on. Take them off." He lay down and was out.

I was again wet and freezing when another Marine shook me. "Sir, it's 0400. You want me to sound reveille?" I sat up, my head in a fog, my mouth cottony.

"Thanks, no, don't worry about it. I'll take care of it." I entered the compound's hut for the first time to wake up my squad leaders. It was at least fifteen degrees warmer in the tiny room filled with Marines. In my head, I cursed the lieutenant still sleeping under my poncho liner. I whispered, "Squad Leaders, get your guys formed up. We are stepping at 0500." I turned to my radio operator, Lance Corporal Eric Rose. "Rose, turn the radio on and let main know we're about to return to base."

Rose went to work, then called me, "Sir, main wants you on the hook." I answered to flames coming from the handset, "ONE ACTUAL! WE ARE ABOUT TO SEND A RECOVERY FORCE FOR YOU!!! WHY THE HELL AREN'T YOU UP ON THE NET?" I was flabbergasted.

"Uhh, negative, sir. I informed the watch we were going dark on communication for two hours to conserve battery power and he approved, sir." He was unimpressed with my answer.

"We'll have a talk when you get back. Out."

The Company CO was at PB Fires when I got back from my first mission. He was not happy with me. Will Donnelly and I both got hauled into the room we used as a combat operations center. He told us both he was going to either relieve us of duty or kick our asses if things kept going the way they had in our first thirty-six hours of operations. I was separately counseled for my "cowboy attitude." I thought of a favorite passage from *With the Old Breed*, "In combat, I saw little, knew little, and understood still less about anything that occurred outside K

3/5. We had our hands full fighting and trying to survive moment to moment."

This was not how I had seen combat going.

FIVE DAYS AFTER our first combat mission, we patrolled back to FOB Inkerman from PB Fires and stayed overnight. A staff sergeant from the company headquarters delivered an interpreter to me in the First Platoon area. The staff sergeant pointed at him and said, "Sir, this is Zainullah Zaki. Call him Zak. He's your translator. He's ready to go."

Like everything else in my life, getting an interpreter was a competition I wanted to win. I wanted First Platoon to have, and to be, the best at everything. Looking at the other interpreters I'd seen, it was clear Zak was heads above them. I thought to myself, "This guy speaks pretty good English and he looks to be in pretty good shape. I need to protect this guy in combat, but also protect the platoon from other people in 3/5 finding out we have a good interpreter." When you're a front-line unit, you're always up against the threat that, if you've got somebody that's really good, someone higher at the company or battalion headquarters will take them. With Zak, I was going to have to play a game of "Oh, this guy kind of sucks, but I guess I'll take him if I have to."

Kilo Company wasn't in a great situation with interpreters, so Zak's arrival appeared to be good luck. In the first few weeks of our deployment to Sangin, most of the interpreters we inherited from 3/7 had already quit or were expected to. It was understandable. Marine unit combat deployments were seven months long. Deployments as an individual augment to another unit, such as an advisory tour, were twelve months. Army unit combat deployments were typically twelve months. Regardless of deployment length, Marines and soldiers went home to the peace of the United States on conclusion. Interpreters were home. Unless they asked to leave their position (or were fired, a rarity), they could continue working for successive Marine and Army units in an endless cycle of combat until they finally quit, were injured, or were killed. Most found alternative employment as interpreters out of combat at the first opportunity. If an interpreter was still there after 3/7 left, it was likely because they didn't speak English effectively, or even worse, didn't speak Pashto.

Initially, I did not know what a huge part of my life Zak would become—how could I? I simply saw ours as a transactional relationship, as I had with the other interpreters. They all seemed to prefer it that way. It limited their obligation to be engaged, and they mostly seemed to enjoy sitting and smoking between missions before joining us just as we left the gate. Zak was another attachment with a capability I needed, like a machine gunner or a sniper. I needed to talk to Afghans, and he was an Afghan interpreter.

Though I was focused on preparing for a combat patrol back to PB Fires, we spoke for a moment. I stuck out my hand to shake his. "Zak! Welcome to First Platoon. We are going to patrol from here at Inkerman to Patrol Base Fires. You ready to get some?"

Zak was polite, almost formal. "Yes, sir, Lieutenant Schueman, thank you, sir." He expressed his concern for my family, such as it was. I asked him about his own and where he was from. I would learn the depth of his concern and love for his family over time.

But ultimately, I was there for the war. I needed a dragon to fight and Sangin District offered it. First Platoon was in a firefight and killed five Taliban within twenty-four hours of arriving at PB Fires. If an interpreter could help me do more of that, great. Zak was coming with us.

MY COMPANY COMMANDER proudly proclaimed his intent to bend the Taliban to his will through a kind of smashmouth counterinsurgency. By crushing their will to fight he intended to turn the people of Sangin away from the Taliban and toward the government. To do that we would patrol as aggressively as possible. We would find and kill Taliban. Sangin lent itself to that. My implementation of Kilo's "smashmouth" strategy seemed too aggressive for the CO so the authors of "shape, clear, hold, build" in Kabul may have been decidedly uncomfortable with the violence of our immediate reality. But Kabul was a long way away, and when someone local is trying to kill you, guidance from people you've never met takes a backseat. My fellow lieutenants and I aggressively implemented the Kilo approach. The Taliban were no less aggressive in their response.

I had been in combat less than a week when, on October 15, 2010, my best friend, Cam West, was wounded by a pressure plate IED that took his right leg, and damaged his right eye, right hand, and his left leg as he was in command of Third Platoon. His radio operator, Lance Corporal James Boelk, was killed outright when he stepped on it. I was under Taliban fire when I heard it over the radio. I put thoughts of Cam away as we counterattacked the ambush position and drove off the Taliban.

As we returned to FOB Inkerman, I was in denial about Cam. It just didn't seem possible that "Big Country" could be hurt by anyone. But Lieutenant Will Donnelly was waiting for me at Inkerman's entry control point and confirmed it.

"Tom, Cam is hit and they don't think he's going to make it."

I was exhausted as I looked at Will. All I knew to say was, "Thanks, man."

Like a robot, I led my platoon to our terrain model, a 3-D rendering of the area of operations in which we could talk about an operation before we did it and what we had done after it was complete. We debriefed our mission. That was the expectation. The Corps is a machine that runs on expectations.

Finishing the debrief, I was on autopilot. I walked out of sight of my platoon and ducked into an empty room we used for meeting with Afghan locals. Alone, I doubled over. I vomited as I sobbed for Cam, Boelk, and all the other men already killed or forever changed by gruesome wounds during the first week we were in Sangin. That night I could not sleep. I was in bed for two days, vomiting and shaking. It was the last time I really felt anything for years.

I admired Cam so much. He was too full of life to be in a coma, trying to hold on to life. Despite both heartbreak and nausea, I had to laugh thinking of him charming my mother by singing "LA Woman" at the top of his lungs, as we all got ready to go out in San Clemente. Even with her cop's suspicious nature she described him as, "gorgeous, charismatic, charming as they come . . . an alpha male type, without being a douchebag."

When she came back out to California from Chicago just before we left for Afghanistan, Cam, my mom, and I ended up having a bonfire at the beach. In the light of the fire, I watched my mom look Cam straight in the eyes and say of me, "You take care of him."

He was uncharacteristically serious in responding. He just drawled a "Yes Ma'am."

BY OCTOBER 18, 2010, Zak had been with us for four days and been in firefights for three of them. He seemed unfazed. He had no formal preparation and no training that I was aware of. Except for being handed a helmet and an armored vest, he was a civilian, no different than any other Afghan. But Zak was clearly there to be there, to absorb all the risk, all the danger that we Marines did. Though we had not had any deep conversations, it was obvious he was moved by more than a paycheck and a spirit of self-preservation. He had what Marines call "fire in the belly." He wanted the same thing we wanted, to go as hard as possible to find and kill Taliban.

The mission on October 18 was for First Platoon to establish a position from which to do just that. We would ambush the Taliban and give them a taste of their own medicine. I directed First Squad to be ready to execute the ambush at 5:00 A.M. Third Squad would conduct a bait patrol at 7:00 A.M. to draw the Taliban into a position in which First Squad could kill them. The day began like many days in Sangin, under fire and counter to the plan.

Third Squad was executing their bait patrol when they came under heavy and accurate fire from four Taliban with PKM machine guns and AK-47s. The squad was separated by an open area and unable to reconnect due to enemy fire. When Lance Corporal Smith saw a Taliban firing from a doorway, he keyholed the doorway with a 40 mm gre-nade, killing the fighter instantly. The enemy ceased firing at us and we moved into a cornfield for concealment. Within a few hundred meters, the enemy began harassing fire, attempting to figure out where we were within the corn. We entered a canal to get relief from the probing fire, but as the last Marines reached the canal, the rear element took very effective fire.

Lance Corporal Syrus Isfahani jumped out of the canal under heavy fire, ran down a path as yet uncleared by the minesweeper, and forced the Taliban to flee from him and his M240G machine gun. As the pa-trol continued down the canal, Corporal Justin McLoud saw an enemy spotter and killed him before we finally linked up with First Squad and sat in the ambush position for about an hour. With no activity, and ex-pecting none in the wake of the firefight we'd already had, I decided to head back to PB Fires.

While on the move, Kilo Company called to tell us the enemy was now in the kill zone of the ambush we'd departed thirty minutes before. I explained that we were no longer there, as I had reported thirty minutes before, and got an order to return to the Taliban's last known position. It was our ambush position. I argued against that idiocy and lost.

As we headed back west in the canal, we were ourselves ambushed in the same kill zone within which we had planned to hit the Taliban. It was insanity. Machine gun fire ripped down the canal, kicking up dirt from the banks. First Platoon Marines jumped out of the canal and began returning fire. Lance Corporal Isfahani once again put himself at risk. With incoming rounds dancing around him he began again laying hate with his M240. Nearby, Lance Corporal Derek Goshke was shot in the leg, leaving him gasping on the canal bank. Our corpsman, Doc Rashad Collins, was separated from Goshke by a hail of bullets bisect-ing the canal. He ran through them to reach and treat Goshke, treating his leg injury while submerged in the fetid irrigation canal. Another corpsman, Doc Long, ran into the fire pelting the long axis of the canal to assist Doc Collins with Goshke, who was cursing the Taliban and hurling obscene insults at them as fast as any Marine was firing.

Staff Sergeant Henley was on the radio calling in a medevac helicopter as I worked up a fire mission from the Kilo Company 60 mm mortars.

"Sledgehammer Main, this is Sledgehammer One, adjust fire, over!" The answer came back, "Sledgehammer One, this is Sledgehammer Main, adjust fire, out!" I provided the location and description of the enemy target and could soon hear mortar rounds whistling as they passed over our heads on the way to the next tree line. As the mortar rounds impacted to our north, I called adjustments, then, "Sledgehammer Main, fire for effect!" The exploding mortars killed four Taliban.

The incoming fire slowed down, but we were still under fire in our own planned kill zone, walked into it by the company headquarters. I asked for air support, but it was slow in coming. When I asked for an update on getting air on station, headquarters told me, "Just sight in and return fire!" I looked at Zak and shook my head in exasperation as I keyed the handset, "No shit return fire! Where's my fucking air?!" Zak looked at me with a sardonic smile and a shoulder shrug that said, "What can you do?" Despite rounds cracking in the air overhead, he was unflappable.

Sergeant Myers took over control of the 60 mm mortars as I turned my focus to getting the platoon out of the canal and back to PB Fires. We started to move, with my RO, Lance Corporal Eric Rose, the first Marine out of the canal. He secured a foothold to cover the rest of the Marines as they exited. Under fire at his position, he held fast, pulling reluctant Marines out of the canal and smacking them on the ass to get them moving. After one more volley, and a corresponding blast of 40 mm grenades from us, the Taliban ran.

In our first nine days of operations, First Platoon had killed fifteen Taliban. We were fighting harder and killing more enemy than any unit in 3/5. By extension, that meant we were fighting and killing more than any unit in Afghanistan. I saw that as in keeping with my CO's direction as I understood it. But pushing that hard can create political and personal issues that people more removed from the sharp end of things perceive as an unnecessary risk. They may even be right, given the clarity that distance can offer.

In my aggressive approach to the area of operations, I followed the rules of engagement, but I was not particularly attuned to political liability that increased the farther anyone got from PB Fires. But I was consciously absorbing a lot of physical risk. It made me short of patience with people I saw as less engaged, which, to be honest, was everyone who was in Afghanistan and not in First Platoon. When I got back to PB Fires that day, the CO told me he was pulling First Platoon back to FOB Inkerman over his concerns that I required closer supervision. We were moving, but thankfully, Zak was coming with us.

ON NOVEMBER 9, 2010, Sergeant Jeffrey Iwatsuru's Second Squad was on patrol when a Marine left the marked path and triggered an IED. The IED explosion "went low order," meaning that it did not fully detonate, a factor with the Taliban's homemade explosives and the wet earth within which they hid them. The explosion still broke the Marine's leg.

Staff Sergeant Henley, Zak, and I launched with First Squad under Sergeant Humphrey as a Quick Reaction Force to support Second Squad's return to base carrying the injured Marine. Unlike the three squads, who rotated, Zak rarely got a break. Zak was usually with whoever was outside the wire. But as the First Platoon leadership team, it was rare for Henley and me to be out together.

Sergeant Humphrey's squad secured a path and established a position from which they could overwatch Second Squad as they reentered friendly lines. As Humphrey's Marines were breaking down their overwatch position to move back to FOB Inkerman, the rear element of the squad was pinned in a canal by heavy and accurate machine gun fire from Taliban hidden in a tree line. I ran to Sergeant Humphrey, shouting, "Let's go, we gotta move on that gun!"

Staff Sergeant Henley and I were converging on Sergeant Humphrey as we rushed to develop a plan to relieve the pressure on the pinned-down fire team from the Taliban machine gun. I looked back to see Zak on my heels. There would be no need for an interpreter where I was going. He just saw me start running and he did the same.

Then Sergeant Humphrey stepped on a pressure-plate IED.

The sound and force enveloped me in sensation. All went black. I have no idea how long I was out, but when I came to, the first person I saw was Zak. He knelt beside me, over me, my rifle in his hand, protecting and supporting me as I fought to my feet. I imagine we looked like a filthy Madonna and child in the moment. As Zak helped me rise, I looked to my left and saw Staff Sergeant Henley. He lay in a heap, a tangle of his own limbs, himself struggling to rise. He had also been knocked out and was shouting at me, unaware that blood poured from his blown-out eardrums. I could make no sense of what he was saying. Then I saw Humphrey and stopped trying to.

When I got to him, Sergeant Humphrey's right foot was completely blown off. All the muscle and tissue on his left calf was gone. His left thigh lay open from a piece of shrapnel. Later I would learn his jaw was broken. But for now, he lay at my feet screaming, "Is my dick gone?! Is my dick gone?!" It is usually the first question a Marine asks after an IED blast. I wish I did not know that. I wish I did not know that if it is gone, you lie and say it isn't. "No, brother, you're fine. Everything is okay there."

Doc Collins and Corporal Matt Bland rushed to Sergeant Humphrey and began treating his grievous wounds. I called in a medevac as I held the meat of his calf to the bone as Nikirk wrapped it. With Corporal Sean Leahy directing the squad's fire upon the enemy, Corporal Spivey swept the kill zone for more IEDs and then swept a clear path to the point at which the medevac aircraft would pick up Humphrey. Through it all, Sergeant Humphrey remained poised, directing his own treatment and movement. He lay upon the litter, apologizing. "I'm sorry. I know I'm too heavy. Just put me down and take a rest." The guy who in a few short moments would be resuscitated in flight, who was probably going to die, just wanted us to put him down because he was worried that he was too heavy for us. In the imminence of his own death, Humphrey was concerned for his brothers. It was the essence of what *Semper Fidelis* truly means.

As we carried Humphrey to the medevac helicopter, all I could think was, "What a bad day." It was insufficient for the moment, but it was all I could think to say. They tried to put me on the helicopter, but with Staff Sergeant Henley and Sergeant Humphrey both wounded and gone, there was no way I was leaving.

When we returned to FOB Inkerman, the CO handed me a towel to wipe the blood off my hands and gear and said, "Sergeant Humphrey is gonna lose a leg, but the surgeon says he will live. But what happened out there?"

I looked at him and exhaled. "I don't know, sir. That file is deleted." There was little else to say, so I turned away and went back to debrief my platoon. I was tired and sad and likely concussed, but I had a job to do. Zak always did his and more; I could do the same. Bringing out the collective narrative in the debrief began to bring back the details for me. But Zak standing over me, ready to kill to protect me, stood out in stark relief from the start.

That vision, Zak silhouetted against dust still swirling from the explosion that flattened me, color and detail becoming clearer as my consciousness returned, is one I still see sometimes as I wake. It says something about Zak as a man, as a battlefield interpreter, and as my friend. It was common to hear Americans say of the Afghans, "We can't want it more than they do," but it often seemed we did. Zak was a huge exception. He was not a soldier, but he was there to fight. He understood our missions in a way that other interpreters did not. Most interpreters were like specialty items in our packs, inert and just riding along until it was time to employ them for their purpose. But Marines are utility players capable of addressing a wide array of circumstances and expected to prevail in each. Like a Marine, Zak was an active member of First Platoon. He did not shy away from danger. He did not question doing whatever the mission demanded, because he was committed to it, for Afghanistan. Zainullah Zaki ran to the sound of the guns in a way that would have made a Marine infantry instructor smile.

Standing at the debrief, talking through what happened as is expected after every operation, I told the platoon that all of us were going out the next day for Humphrey's revenge. I thanked Zak for his willingness to always be where he might be needed, particularly when it was protecting my life. I had already come to see Zak as more than an interpreter, but now there was no way to call him anything but a brother.

As a whole, 3/5 had taken a significant number of casualties in that first six weeks of the deployment. Several adjacent units had been hit hard. Sergeant Humphrey was the second casualty in First Platoon, but the extent of his injuries was in sharp contrast to Lance Corporal Goshke's through-and-through leg wound on October 18. Moreover, some men are more than the space they occupy.

Sergeant Humphrey was the Swiss Army knife of the platoon, appropriate for use in any situation. Losing him left a gap in our collective capability. Humphrey's wounding also had the potential to create a crisis in confidence. Amputation leaves a lot of space to consider for young men with years ahead of them. Men begin to calculate their own odds. As a leader, responsible for my men's welfare, I could not help but wonder what I could have done to protect him from the injury. Some Marines seem untouchable and when the enemy proves differently, less capable Marines begin to question themselves. But I could only do so much to head that off with my words. That was a concern best addressed through action, by getting back outside the wire the next day. To prepare for that, I needed to rest.

Leaving the Marines to decompress in their own ways, I went to my room and wrapped myself in my sleeping bag while watching *Lonesome Dove* on my computer. Reason did not play into the sense of culpability I felt for Humphrey's wounds. That was a result of conditioning. A Marine leader is responsible for all that his or her unit does or fails to do. By extension, I was responsible for anything that happened to them. The accumulation of all of it weighed on me.

That command is lonely is a trite observation, but it is no less true for the countless times it's been observed. It should be paired with the fact that leadership hurts. Every time you discipline a Marine that you love, every time that you watch them bleed, every time that you watch them breathe their last, it hurts. I was twenty-four years old and living a reality that I now look back upon and wonder if some of it even truly happened. I want to deny that humans could be as callous as I became in the destruction of people put on this earth by the same God I was. I was convinced I would not survive Sangin, but I had realized that life is a gift and if I did survive I had some growing up to do.

I prayed to God and asked that if I did survive, He would help me live a grateful, meaningful life that would do honor to the people denied that chance. I was largely incapable of making any objective decisions myself beyond where to patrol next and how to fight through the next ambush, but I was ready to acknowledge I was not the man I professed to be. I called Andrea and told her I wanted to be a better man; that I could only do that with her beside me. I apologized for mistreating the woman who loved me and begged her to take me back. She agreed. Though I still thought it unlikely I would live to see her again, the idea was something to live for.

At some point that evening, the CO came in and told me that my friend and mentor Lieutenant Robert Kelly was dead, struck and killed instantly by an IED earlier that day. I rolled over and stared at the wall. I thought about Rob, Cam West, and me spending the last day before we deployed drinking beer at Del Mar with Robert and his family. I thought about an MEU and a trip to Japan. Then I went back to *Lonesome Dove*.

The next day was the 245th birthday of the United States Marine Corps. First Platoon killed four Taliban. It didn't bring anyone back.

After enough repetitions, you get a sense for when bad news is coming. You feel it before you know why, like you just walked in on your parents fighting, a state at the center of my earliest memories. They stop long enough to look at one another, trying to decide whether they are going to play it off to preserve the fiction of a happy family or whether today is the day you get to learn the truth. It's a pregnant moment that turns the air electric, then something tips toward the new reality and you can either try to keep up or get knocked down and dragged by the rush of resulting events.

On November 25, 2010, I walked into the Kilo Company combat operations center and deduced from conversation that Will Donnelly was dead. It was Thanksgiving. Will got married on September 11, 2010. Two months later, on Thanksgiving, he was shot through the forehead and killed. I took a moment and thought of the friend I had known, the guy I lived with in San Clemente; the guy who had taken on the duty to tell me about Cam. Then I turned in another patrol overlay and got First Platoon prepared to depart on another mission.

BY THANKSGIVING 2010, most of my closest friends in 3/5—Rob Kelly, Cameron West, Will Donnelly, and Gunnery Sergeant Carlisle—had been killed or wounded. Only Staff Sergeant Henley and Lieutenant Vic Garcia, who replaced Cam at PB Fires with Third Platoon, remained. I only saw Vic in passing. Another IOC classmate, James Byler, had been horribly wounded, losing both his legs. The death or grievous injury of friends with whom I had been living, training, and fighting for two years left me feeling dead inside. I was just waiting for my body to catch up.

We had arrived in Sangin in October 2010 and settled into a life of brutal sameness. We patrolled through suffocating heat and penetrating cold, sometimes multiple hours-long patrols in a day. We engaged the elders. We did the post-blast analyses. We responded to the ambushes. We called in the medevacs. We called in the air support. We zipped our friends into bags for a final ride home. They were men I loved because I knew them or I loved despite not knowing them, for the simple fact that they had volunteered to serve as Marine infantrymen in a time when that meant war. But it was harder and harder to feel that emotion as the toll of dead and wounded grew and I felt increasingly ground under the wheels of the war.

When I look at the log I kept between October 2010 and May 2011, I am drawn back to a time rendered in the contrasting browns of desert and the vibrant greens and reds of poppy fields. The log's shorthand reveals a stunning background staccato of violence that, had any single day of it happened in America, would have attracted an army of therapists, reporters, and editorialists clucking and gasping as political theater. For us, it was just a constant desensitization to, and normalization of, the savage and bizarre. Among young men left unchecked, killing almost becomes an end unto itself. War is nothing but waste at its essence. Over time, our efforts in Sangin almost ritualized that waste. My entries give lie to the notion of glory in war, but at the time I was too close to the fire to worry about the burn.

"Second Squad conducts contact patrol at 1100. Patrol takes accurate fire from two enemy fighters engaging from murder holes. Sergeant Iwatsuru takes bold and decisive action to maneuver on enemy firing points. He appoints Corporal Orozco's team as a support by fire while he rapidly closes on enemy pos with Corporal Sotelo's team. Lance Corporal Isfahani begins taking accurate fire at his support by fire position. Sergeant Iwatsuru's aggressive action forces enemy to retreat. Second Squad goes firm and I coordinate Close Air Support. Close Air Support fires, killing four enemy fighters. Second Squad can hear Taliban yelling in pain and decides to maneuver on enemy casualty evacuation. As Second Squad comes on scene we observe the enemy in complete disarray. We are able to identify two enemy fighters and Corporal Orozco and Lance Corporal Rose kill them.

"Sergeant Humphrey's squad discovered five pressure plates on a rooftop. As the squad began searching the compound they came under heavy fire. The enemy initiated the ambush from the north with an RPG shot and then began laying down a heavy volume of fire from a Medium Machine Gun in an elevated position to the squads south. As the squad began maneuvering towards the RPG team they encountered a pressure plate IED in a canal. The enemy was attempting to force the Marines towards the 25-pound charge of the IED.

"Lance Corporal Isfahani observes two IED emplacers at TB check point. I engage with an M203, blowing one of the guys to pieces. Sergeant Bland kills the other one with his Mk12."

As the months wore on, the losses mounted, eventually more than any battalion in the twenty-year war. Sadness became simple numbness. But there were days that punched through that. I had few people with whom I could share those moments. Staff Sergeant Henley was my platoon sergeant and the foundation upon which the platoon stood, but there were things he needed to not hear from me. There's a fiction about fearlessness and invulnerability that must be maintained between a Marine leader and his or her men. It is as much a part of the leader's duty as any other element. Honesty is fine to a point, but the led must believe that the leader is committed and willing to bear the burdens of the missions he asks them to undertake, that their potential sacrifices will be honored.

Though I thought as much of Zak as I did a Marine, I did not have that same duty. I could discuss anything with him. By his continual presence next to me, Zak had become my friend and confidant. He was not a Marine, so I could share thoughts with him I could not easily expose to men I led, who needed me to be strong at all times. Both under fire, and sitting by one another, he became my brother.

Tactically and professionally, Zak was the constant I needed. He and I patrolled with all my squads, he even more than I, and could tell me things about squad dynamics that happened outside my view. As we walked and talked with local villagers and patrolled with members of the Afghan National Army, Zak served as the interlocutor for it all. I dictated the tempo and the energy of those moments but Zak mirrored my energy, whether questioning people in villages or when the bullets were close. If there were advances made in selling our side of the story to the people of Sangin, the victories lay with Zak.

Because we were working in parallel, because he was literally charged with conveying both my words and their intent, we became almost one person in a sense. The closeness of our working relationship made it natural that he became a sounding board and someone to whom I could unburden myself about cultural frustrations. Soon it was natural that he was my friend.

At Inkerman, Zak and I both had our own rooms in which we could talk freely. That mattered. On return from missions, I had to wear a mask in front of the Marines during debriefs. I had to project calm, to kind of laugh at the death, and joke about how close we were. But in private, I had a different version of those conversations with Zak. Zak was always physically closest to me on missions and experienced things similarly, if not identically. We each decompressed by freely talking through what had just happened, a post-debrief debrief. We had a brother kind of relationship in which I could trust him because he was physically with me in all the darkest moments. Because our conversations were limited by the extent of Zak's English, we could reach the very essence of an issue, without a Marine's pretense.

"That was a bad firefight today, right?"

"Yeah! I was afraid I would die!" He always accompanied serious statements with a laugh that rippled up from his belly. His honesty opened the door for my own. I could talk to Zak about my faith and my failings within.

"Zak, I have not always been the man I want to be." I could address my deep sadness over the loss of my friends or my desire to be a better man for my once-again girlfriend, Andrea.

Zak often talked about his brothers and sisters. He had concern, even then, for his father's health. I told him what it was like growing up with my parents so far apart. That I had rarely seen my father was sad to him. It was sad to me, too. Zak's desire for a university education was evident when he asked me about Loyola and the things I studied there.

Sometimes it was just the simple support of a friend when I needed one. "Okay, I just got chewed out again."

"For what you got chewed out?"

"Some bullshit one of these people that never leaves the wire didn't like."

"Hey, sir. Fuck it."

In a place where unrelenting violence was stealing them from me, Zak was my friend in all the meanings of the word.

9

A Brotherhood Born

ZAK

IT HAD ALREADY been a long mission the day Sergeant Humphrey lost his leg and Staff Sergeant Henley and Lieutenant Schueman got blown up by an IED.

We had been to search a village where the Taliban hid weapons and explosives. The Afghan National Army soldiers on patrol with us found yellow jugs full of homemade explosive, made with fertilizer and aluminum, in the mosque there. There was nothing much we could do when we found the bombs besides destroy them. The people in the village always said the same thing: "We don't know anything about that. The Taliban stops here. We don't have weapons or anything to fight them. If they come, we cannot tell them, Don't do that. If we say anything, they're going to kill us." Sometimes the people were telling the truth. Sometimes they were hiding the Taliban. That day, I told Lieutenant Schueman the villagers were bad people and working for the Taliban, but there was nothing to be done about it because there was no proof.

We had left from searching the village when a Marine hit a bomb. I had seen many Marines injured in just one month. He was lucky and it did not explode fully. We were returning to FOB Inkerman when we got ambushed. I knew that Lieutenant Schueman would run to where the fight was happening. My place was with him. When he shouted, "Let's go!" I followed.

I was behind Lieutenant Schueman when the bomb hit Sergeant Humphrey. I could not see him. He was in a cloud of dirt and smoke. I saw the Lieutenant and Staff Sergeant Henley knocked down. I felt dirt and rocks hit my face, and my ears were ringing, but I was far enough behind the three men in front of me that I was still standing.

I had been with First Platoon for a month by then. They were my team, especially Lieutenant Schueman and Staff Sergeant Henley. I looked at them. They lay on the hard-packed path, not moving. I still could not see Sergeant Humphrey through the smoke and dust. I was afraid they were all dead. But my place was with Lieutenant Schueman, so I started running toward him.

When I got to him, he was not awake. His rifle lay on the ground, just out of his hand. Staff Sergeant Henley was knocked out, too, but he was already trying to sit up. I did not have medical training, so I could not take care of either of them. Henley was shouting, but I could not understand him. Interpreters were not allowed to have weapons, but I thought about all the things the Marines accepted on behalf of my country. I had to accept the same things. There were Taliban attacking us, so I picked up Lieutenant Schueman's rifle and knelt by him to protect him, looking across a field for Taliban to kill. I wasn't supposed to carry a weapon, but I am an Afghan. We all grew up at war. I knew the AK-47 better than the Marine's M4, but they had shown me how to use it for days like this. Protecting Lieutenant Schueman was all I knew to do for him. I knew he would do it for me.

When the Lieutenant became awake, he tried to get up. He struggled, so I got behind him and lifted him. Then he stood up and ran to Humphrey. I followed with his rifle. All I could do was protect him and my team while they tried to save Sergeant Humphrey's life.

When the helicopter came, I said, "Lieutenant Tom, you must go." But he would not. He put Sergeant Humphrey and Staff Sergeant Henley on the helicopter and led us back to FOB Inkerman. He always led us. I was so glad he was alive—he was my friend and I could count on him to take care of me. I was prepared to die beside him. There were so many times I thought I would.

WHEN I APPLIED for the job of interpreter back at Topchi, it was impossible for me to truly understand the things I would have to do. In Kunar, the Taliban and Al Qaeda were attacking the Americans, but a lot of the fighting was in the valleys with the Highlanders, in places like Nangalam. In Sangin, the Taliban was everywhere. Every day was frightening. Every place we went I wondered if I would die, or if my friends, my brothers, would. No matter what, I told my family I was safe; I did not tell them what I was doing. I told them there were no combat missions. My family did not want me in battle, so I decided to keep the truth of my job quiet until I was back in Asadabad someday.

First Platoon was kind to me—we worked and lived together as brothers. We took care of each other. I got local food for them. I told them things about Afghanistan; about our culture, our religion, and the many different people who lived in our country. One of the Marines came from Mexico and became an American. He worked in the fields in California, where he grew up. Surrounded by cornfields, we talked about farming.

"Growing up, we picked fruit in the fields, Zak. Sometimes we picked almonds and pistachios. We were poor, man. My mom sent me to America young to live with my Tia and Tio. We did whatever we could do to make a dollar, man."

I told him about my family cutting our flowers to sell and also about other crops raised in the country. "Afghanistan has beautiful almonds and pistachios in Kandahar."

"Well, why don't you get some of these poppy farmers to flip it and grow pistachio, dog?"

"Why don't you get a job in Kandahar, Lance Corporal?" We liked to tease each other but we could also be serious.

"For real though, Zak, I joined the Corps to get out of the fields. I got my citizenship. I have a car, I got college money, I got opportunities, man. I am proud of how hard my parents worked to get me to America so I could be something."

"I wish we could have these things in Afghanistan. It is why I am here now."

"It's why I am here, too, dog. I got your back, bro." He reached out to bump my fist. I was proud to fight with men like him.

We were barely men; some of us were still teenagers. They told me about growing up in America, about football games and dances where girls and boys danced together. They showed me pictures of themselves in suits with girls in dresses that showed their shoulders and more. I told them that would never happen in Asadabad, but that I hoped we could at least get good schools for girls.

Some of them were hunters and fishermen like me. I told them about building rafts out of tire inner tubes and floating them across the Kunar River to the island where we played. I told them how we would take a goat on the raft and slaughter it on the island to eat with the sweet, white flesh of the fish we caught. Sometimes we shot birds with a shotgun and cooked them, too.

We compared our countries and I told them how I hoped that someday Afghanistan would be more like the United States. They told me the good and bad things they saw about America. We all wanted the best for each other and our countries.

The Marines of First Platoon helped me out with everything I needed to survive in Sangin. When their families sent them packages, they gave me things from them, too. The Marines only cared about how I performed in the fighting. The more I went on patrol, the more I was accepted as one of them. I did not want to let them down. A corporal said to me, "Before you, a lot of interpreters came here and were crying. They didn't want to fight, they didn't want to go with us on missions, they were scared, so they quit." I didn't have a gun, but it was my fight as well.

Before we went on the patrols, we had a mission briefing. Many interpreters did not attend them; they did not consider themselves part of a platoon, so they showed up at the last minute for missions. I made sure I understood everything about the mission before I went outside the wire. First Platoon was my team. In Sangin, they were my family.

Sometimes Lieutenant Schueman stayed at FOB Inkerman and I went on patrol with one of the squad leaders. I liked to go with Sergeant Decker. He was a brave man and a good person to work with. There was a Japanese American, Sergeant Iwatsuru. He was also a kind person. From him I learned that his family came from Japan to America a long time ago and became Americans. To me that was an amazing thing. In Afghanistan you are your tribe, your ethnic group. You cannot become something else. In America, you can become American.

In the 1960s, thirty years before the Taliban, Kabul was a growing city with modern buildings and Western freedoms. The Soviet war and the Taliban rule that followed destroyed our country. I knew none of that as a child. I had no way to learn it under the Taliban. But since the West had come to Afghanistan, I had access to information. Now I envisioned a country where my mother and sisters could vote or walk by themselves. I wanted our people to write what we wanted, say what we wanted, and believe what we wanted without getting reported and killed for apostasy. I wanted Pashtun and Hazara and Sadat Afghans to have a government that could represent all of them.

I was learning from the Marines about America and Americans. They had many things I admired, but my heart was in Afghanistan. I was proud to be working with my U.S. partners for a good security situation in Afghanistan. I was proud to be a part of the American mission and my country's future. It was important work that I thought would mean a better life for millions of people. I still believe that.

It was not always easy to focus on the bigger picture. We went to a new village every day and searched for weapons and explosives. Lieutenant Schueman and I visited the village elders often. I interacted with the people in the villages all the time. I told them things that the Marines and the government would do for them: roads, schools, markets. The people in Sangin did not seem to want what we were offering. They did not feel a connection to Kabul. They wanted to farm and be left alone. They were too scared of the Taliban. Or they were the Taliban. We knew bad guys were in the village—in many cases, they ran the village.

We didn't do operations at night in Sangin because it was too hard to find IEDs without hitting them. We were very aggressive during the day, but because we were not patrolling at night, the people never fully trusted us. We were all playing a game. The Americans wanted to improve security. The villagers promised change if they did. The Taliban disrupted security. The Americans asked the villagers why they weren't doing more to help make security better. The villagers said they could not because of the Taliban. They knew the Taliban would still come, no matter what we did in the day. We did not own the night; that belonged to the Taliban.

It was frustrating for me and for Lieutenant Schueman, but more for him. Lieutenant Schueman wanted to win. I understood the different parts of Afghanistan and the cultures in those places. I just wanted to not lose. That would be enough to make a better Afghanistan. The history of our country has always been about who could endure the most. I knew that the people in Sangin were not going to change until the Taliban were defeated because they did not believe they would survive if they did not do what the Taliban said. The war had to happen to build the kind of country I was fighting for.

I tried to help the Marines convince the villagers that they were there to help. Sanctity of the home is very important in Afghanistan. I helped Lieutenant Schueman understand Afghan culture, so that if we could not make things better, we would not make them worse. Lieutenant Schueman listened to me in the villages when I said things like, "The man of the house isn't home right now, we should come back later." To have searched a home when a man was not present could have placed a woman in danger. It would at best be a sign of disrespect. There could be rumors that she had been raped, a dishonor for the family that is sometimes solved in Pashtun culture by the "honor killing" of the girl or woman. He understood these things when I told him, and if I did not have time to explain, he trusted my judgment.

I also helped him work with and understand the Afghan National Army soldiers at FOB Inkerman. Afghan culture is not Marine culture. The Marines and the ANA did not always do things the same way. But we were in Afghanistan, not America, and I knew that someday the Marines would be gone and the ANA would need to be ready to fight the Taliban alone. I talked to the ANA at FOB Inkerman even more than I spoke to the people in the villages. Later, during our conversations, I explained to Lieutenant Schueman what they thought and said.

The ANA did not see a purpose in patrolling every day. They knew that the people of Sangin would not kick out the Taliban. Some families in Afghanistan had sons in both. In Sangin, more were Taliban. The Afghan soldiers grew up under the Taliban, and they did not want them back. But mostly they wanted to finish their conscription alive and go home. Many of the soldiers only served to have a job. They did not share my ideal of a better, freer Afghanistan. They just wanted a little food, maybe some hashish, and to be left alone. Afghan corruption affected them a lot; their pay and their will to fight suffered. They just wanted to get what they could while they could.

Many times the ANA did not want to patrol when the Marines wanted them to. Most of them were not aggressive. They did not want to get in firefights. Many of the Marines seemed to enjoy it. The ANA knew that they would be in Sangin after the Marines left, and they understood how the villagers felt about a central government. Many ANA soldiers did not want to die or be injured for people who did not care about their mission. The Marines told me about the U.S. Department of Veterans Affairs. There was no VA in Afghanistan and no plan to take care of injured Afghan soldiers. The wounds that they saw Marines get were not survivable in the Afghan medical system. When the Lieutenant wanted the ANA to do things in ways they were not used to, or they did not want to do, it was important to manage how they heard his words.

Lieutenant Schueman and I were together every day. I understood how he thought. I was very proud when he said, "Zak, you're very effective at not wasting my time, or anybody else's. You know what my intent is and what I hope to achieve, so just have the conversation on my behalf." I was proud that he and First Platoon trusted me to do what needed to be done.

Still, there were brave ANA soldiers. Lieutenant Schueman wanted to know how to get those ANA to go on operations with us. I spent as much time as I could with the ANA who shared my ideas for the future. We sat and ate the Afghan food that I missed while living with the Americans. The ANA invited Lieutenant Schueman to eat with them and he always said yes. That helped. Respect is important everywhere, but especially in Afghanistan. The more the ANA knew that Lieutenant Schueman respected them, the more likely they were to accompany him on operations. If the Afghan government had treated the whole ANA the way Lieutenant Schueman treated the good soldiers at FOB Inkerman, Afghanistan would have been better off.

I began to enjoy some of the missions. It felt good to be part of a team and care about people who cared about me and my country's future. Kilo Company wanted to make Highway 611, running north to south through all of Sangin, secure for people to get to the market in the district center. To do that, we had to stop the Taliban from putting in IEDs on the road. First Platoon cleared the area along Highway 611, securing all the high ground along the road. The British tried to do the same but the Taliban attacked them with rockets and grenades on timers. The Marines destroyed almost all the old houses for miles along the highway with huge lines of explosive charges deployed with rockets that dragged the charges out so the Marines could blast clear long stretches of ground. Before it happened, First Platoon went out and I paid the owners and warned people of what was going to happen.

Most of the houses were rubble after the explosive charges, but we took over one of the old buildings still standing as an observation post (OP) along the highway so we could watch for Taliban placing IEDs. It was a house the Taliban had lived in before. They left their homes when the British came, but they still hid in them to attack American convoys going from Lashkar Gah in the south to Kajaki Dam at the northern end of Sangin.

We had just occupied the house when I had to go to the bathroom. I went in a room in the compound to use the WAG bag. The Taliban ambushers had used it for a bathroom, too. There was poop everywhere. They just went on the floor and left it. There was also a pile of branches in the room. I thought, "I should search that pile to make sure there are no bombs before I poop." When I picked up the branches, there was an RPG-7 rocket-propelled grenade launcher. I took it to Lieutenant Schueman. He, and all the Marines, were so happy because that was one less launcher the Taliban could use to attack convoys.

The next day, the Marines were building a bunker on the roof of the OP. The roof collapsed and they fell into the same room where I found the RPG. They were covered in Taliban shit. It was funny then. Now I think it was maybe a sign of my country's future.

Mission after mission, I felt lucky to be alive. There were so many IEDs, I thought I would have to hit one eventually. We found most of the IEDs with our eyes, not the metal detector. Maybe we saw something the Taliban placed as a sign to each other. I was afraid to die from an IED, but I could not refuse a mission in my own country that men from America accepted. What kind of man allows another to do his job? What kind of man expects someone from another country to die for his own? Lieutenant Schueman and his Marines were in Afghanistan to help me and my family. Nothing the Taliban did stopped them. I felt that I owed them everything I had.

Many times, the Taliban shot at us to try to push us into the places where they hid IEDs. On a day we patrolled out from the Highway 611 OP, we were ambushed. I took cover in a canal along with two machine gunners, Espinoza and one whose name I forget. We looked out at where the Taliban were firing at us from behind trees on the side of the canal. Lieutenant Schueman called in helicopters to come and help us attack the Taliban.

"Gunrunner, this is Sledgehammer One, we are troops in contact, requesting immediate close air support, over."

"Roger, Sledgehammer, this is Gunrunner, whaddya got?"

"Gunrunner, our location is in the east–west canal, by the footbridge. Enemy location is three hundred meters north of my pos, in the next east–west tree line, at the northeast corner of the cornfield. I will mark with 40-millimeter."

"Roger, Sledgehammer."

I watched the helicopter come in. He fired rockets at the Taliban. Then the gun on the front of the helicopter started firing. It sounded like a giant chain saw. I could see the ground torn apart. Espinoza was shouting, "Yeah, motherfuckers! You like that?!" Firing stopped from the Taliban ambush position. Espinoza and I began to discuss whether the Taliban had escaped or were killed. I thought, "How could anyone survive that?" But the Taliban always came back.

Eventually, Lieutenant Schueman told us it was time to move. When I stood up, I saw that the three of us were lying on top of two IEDs. I felt the ground, then I saw some wires. I looked at the machine gunners and said, "Don't move! There's IEDs!"

I called to Lieutenant Schueman, "We need the minesweeper!"

The minesweeper walked slowly to us. He was sweeping and probing the earth as he walked. He found the pressure plates of the IEDs where I had been lying. When we moved, he found the machine gunners had been laying on three explosive charges. The hole was huge when the minesweeper blew them up. There was nothing to do but laugh. That was just another day in Sangin.

Even though the Marines became my brothers, I missed my family. For me, every day in Sangin seemed the same, but my family still had lives in Kunar. While I was in the OP on the highway, my brother called to say he graduated from university with a degree in English literature. Another brother graduated from high school. I was sad to miss the celebration so we had our own celebration in that old Taliban house. I went to a house nearby with a Marine and bought a goat. One of the Marines butchered and cooked it. We ate fresh meat for the first time in months.

It made me more homesick but it meant a lot that my Marine brothers celebrated my real brother.

A FEW MONTHS into my deployment I got leave and went back to Asadabad for two weeks. I took a helicopter to Camp Leatherneck. From there I was on another big plane back to Kabul. I got a taxi to Nangarhar and then on to Kunar and Asadabad. The whole time, I was afraid of Taliban checkpoints, so I hid my interpreter ID in my shoe in case we ran into them. Usually they were only out during the night, so I did not see them.

When I got home it was nice to just spend time doing the daily routine of the house. There were no IEDs and no one was shooting at me. I felt calmer each day I split wood and took care of the animals and the crops. I still heard shooting inside my head, though. I slept as much as I could, but that came with bad dreams about patrolling in Sangin. I told my parents nothing about my dreams or my job, just that I helped the Americans speak to our fellow Pashtuns to help them stay safe and have better lives. Too soon, it was time to return to Sangin.

Many interpreters never came back from leave. It was hard going back to Sangin, but I had a duty to Lieutenant Schueman, First Platoon, and my country, so I said good-bye to my blood family again and started the journey back to my Marine family. When I got back, the Marines were happy to see me. But I had missed some time. When you are in war, time moves very fast. Things change faster than in normal life. I had to catch up, so I went right out on patrol.

On November 24, 2010, Lieutenant Schueman and I were with part of First Platoon watching over another part as they moved through some compounds. We were ready to provide fire if the Taliban attacked them. Usually Lieutenant Schueman and I were with the team moving. But that day we were in the overwatch position.

A Marine named Lance Corporal Arden J. Buenagua found an IED. Buenagua was very good at finding IEDs. He usually operated the minesweeper looking for IEDs at the front of the patrol. It was dangerous work and he was brave. The IED he found was a big one, an artillery shell. It was connected to a pressure-plate trigger made so it completed a circuit and blew up when a Marine stepped on it. He had found many before that one. While he was destroying that one, other Marines searching a compound found three more.

Lance Corporal Buenagua went inside the compound to check the three IEDs they found. He identified the first one as another artillery shell. While he was looking at the second one, the first one blew up. He was torn apart and killed instantly. Lieutenant Schueman and I could not move from where we were. We heard the explosion and watched the cloud of smoke and dust rise into the sky from the compound. We could only listen on the radio as Marines were doing what they had to do.

"Zak, I feel so helpless right now," Lieutenant Schueman said. I put my hand on his shoulder and we watched and listened.

The company gunnery sergeant, Gunnery Sergeant Carlisle, was with Buenagua. He called for a medical evacuation helicopter. He also asked for the Quick Reaction Force to bring more Marines from FOB Inkerman. He wanted them to carry Lance Corporal Buenagua out in a body bag because of the condition of his body. As the Marines came from Inkerman, the Taliban ambushed them. I watched the Lieutenant's face. I could see he wanted to run to them.

Gunnery Sergeant Carlisle called to us on the radio, "I'm going to take some of these Marines to break that ambush!"

Lieutenant Schueman answered back, "Roger, Gunny. Sledgehammer One Actual is standing by to support."

Gunnery Sergeant Carlisle and the Marines with him left the compound where Buenagua was killed to attack the Taliban. One of them stepped on another IED.

When we heard the blast, I looked at the Lieutenant. I could see on his face how upset he was. He was used to being where the problems happened. Now he was stuck in the overwatch position. We both were. There was nothing he could do but call them over the radio from where he was. No one answered. He called again. Nothing. Lieutenant Schueman was praying out loud, "Please, Lord, don't let them all be dead." I offered a prayer to Allah asking for the same thing.

Then the radio came on. Corporal Leahy, who had replaced Sergeant Humphrey as the First Squad leader when he had been wounded so badly, called to the Lieutenant, "Sledgehammer One Actual, I have three Whiskey India Alpha, one urgent surgical, two priority." He was telling us he had three wounded Marines. Lieutenant Schueman was already on the radio calling for the medical evacuation to come faster and to be ready for more wounded Marines.

Gunnery Sergeant Carlisle called him on the radio, "Sir, I'm hit."

"I know, Gunny, I'm working your medevac. You're going to be okay, brother." Lieutenant Schueman called back.

Gunnery Sergeant Carlisle was already injured and the Taliban were shooting at him but he was on the radio helping Lieutenant Schueman with the medevac and organizing the Marines to attack the Taliban. Corporal Leahy told us later that he had to grab him to force him to lie down and receive treatment.

That was the day I understood United States Marines. They are normal people who want to be something more. Because they want to be special, and because they believe they can be, they are. The Marines are people who never quit, simply because they promise to always be faithful. That was the day I understood Tom Schueman and how much he cares for people. He never stopped working. He never stopped trying to make the situation better. It was the day I knew he would always be there for me if I needed him.

10

The Golf Course

TOM

I LED FIRST Platoon for seven months in Sangin before a new lieutenant and his platoon from another infantry battalion arrived to replace
us. From the time we arrived in October 2010, until January 20, 2011,
when Sergeant Jason Amores became the last 3/5 Marine killed during
our deployment, our time in Sangin was unrelentingly violent. Twenty-
five Darkhorse Marines died during those three and half months. One
hundred eighty-four were wounded; thirty-four of that number lost at
least one limb. The dead and wounded represented 20 percent of the
battalion's strength. The butcher's bill rose to a point that Secretary of
Defense Robert Gates asked the Commandant of the Marine Corps,
General James F. Amos, whether 3/5 should be pulled out of Sangin.
The Commandant, who visited us there, replied, "Absolutely not, we
don't do business that way. You would have broken the spirit of that
battalion."

The Commandant was right. First Platoon fought so hard to subdue
the Taliban, to pull us out would have been even worse than leaving
us there ended up being. First Platoon was in a firefight almost every
day between October 9 and December 28, 2010. We killed twenty-five
Taliban for every Marine the platoon lost, the last of whom was Corporal
Tevan Nguyen, killed by a remote-control IED on December 28, 2010.

The ferocity of the fighting wore us down, but it broke the Taliban's will to go head-to-head with us.

Our combat operations also impacted the ability of farmers to bring in the poppy crop by which the Taliban funded their own. When the Taliban told the farmers they were behind on poppy production, Sangin's farmers said they could not work, for fear of getting killed in a crossfire. With a credible threat to their funding, the Taliban came to the bargaining table. But they came with their fingers crossed behind their back. They offered, and signed, a Christmas treaty saying they were sick of fighting and wanted to join the government of the Islamic Republic of Afghanistan, clean up the IEDs, turn in their weapons, and farm. Up to that point, we had found hundreds of IEDs. The treaty required that we stay on our base for several days and allow the Taliban to clear hundreds of remaining IEDs. Instead, they used the time to plant hundreds more.

When we resumed patrolling after the New Year, we were not in regular firefights. From that perspective, things were more peaceful. But unsurprisingly, given how many were now in the area, IED strikes and finds increased. Fortunately, with the exception of three Marines wounded by IEDs on January 22, 2011, they had little effect on men who had become adept at perceiving visual indicators: the disturbed earth, the bent branches, the pair of rocks that said, "The Taliban messed up the baseline of the environment."

In contrast to a deployment that started in October with an ambush before my platoon was wholly outside PB Fires, by February I could walk to the Helmand River and leave my rifle in the base because no one would fire a shot at me and a rifle is of no use against an IED. Some of that was what we accomplished and some of that was a Taliban tactical choice to avoid gunfights with us. Our adrenaline meters were still calibrated to "daily gunfight," which made February and March 2011 deeply boring. Regardless, with IEDs so prevalent, there was no way to let down our guard.

The diminished pace of operations gave us the ability to focus on some of the counterinsurgency tactics I arrived believing I would employ. I went to Sangin confident that I was delivering freedom, democracy, and liberty like a pizza. My belief that I was going to work hand-in-hand with a mullah lasted fifteen minutes into my first patrol on October 9, 2010.

But even in January I was still drinking the Kool-Aid that there was going to be an awakening like had occurred in Al Anbar Province of Iraq when the locals turned on Al Qaeda. Our higher leadership certainly seemed confident. Major General Richard Mills, the Marine in charge of Helmand, said in February 2011 that the Taliban had "lost the support of the people within the province." That statement did not reflect anything I saw in Sangin and it seemed even more discordant when President Obama announced we would withdraw on a timeline rather than based upon conditions on the ground.

March was stunningly boring. Maybe that was a good thing. Maybe it allowed time for each of us to begin to become humans again before rejoining our families. But the slowdown also gave Marines time to think. As we each had time to reflect on the preceding months and the friends we lost, questions of purpose naturally arose.

It had become tough to verbalize why we were in Sangin. I began to wonder if anyone knew. In my conversations with Zak and the Marines in First Platoon I took a more general approach, one couched in our constitutional duties: "We are here because that's where the Marine Corps sent us. We are here because the nation said that we needed to be." I think if I had tried to have a deeply philosophical conversation, I would have lost my mind, so I kept it tactical. "We're doing what we're doing because there are bad guys that need to be dead. And we're Marines, and we're going to go arrange that."

I still had my doubts. I knew what I thought I had fought for. I was figuring out whether that held up. But I was also thinking about my own return. I knew I would get a different assignment within the battalion. That was normal, though the thought of leaving First Platoon was difficult. I also had every intention of making up for my idiocy with Andrea before I left California. That was life-changing. Amid processing the insanity of the deployment, I had things to look forward to at home. So did my Marines. That made a huge difference for us.

I wanted combat and I got it. Now I wanted peace. All I had to do was make it back home. Zak was home. He wanted the same things I did. I knew, because we discussed it. He wanted an Afghanistan where he enjoyed security, prosperity, and liberty. He wanted an Afghan version of the American dream, by which he could pray, work, and raise a family in peace. He deserved it. But was he going to get it? I had not seen much from the ANA that gave me hope they had the same commitment he did and could ensure his dream held. I was worried for Zak's future. Regardless, we were leaving.

ON APRIL 5, 2011, we were handing our reality to someone else as part of the process known as a Relief in Place/Transfer of Authority (RIP/TOA). RIP/TOA is one of the most dangerous times in a combat deployment. A new unit arrives to replace one that has been fighting for seven months. Over a couple of weeks, Marines who very much want to leave but who also have pride of ownership in an area try to arm the incoming Marines for the experience they desperately want but also dread gaining. It begins with patrols in which leaders from the new unit accompany the departing unit. Over the course of the "RIP," patrols become a half-and-half split between old and new units, then winnow further until only a few of the departing leaders accompany patrols made up of the new unit. Eventually the final handshake occurs and with a "Good luck!" the new unit is responsible for the area of operations.

Demographically, Marine infantrymen are the youngest members of the four services, typically between nineteen and twenty-three years old. That youthfulness increases the excitement of arriving Marines, a factor heightened by the aggression valued and encouraged by the Corps. The new Marines are coming off months of training and certifications for this very event; they are full of ideas for exactly how they mean to influence the area of operations to which they've been assigned. They are technically and tactically correct per doctrine and theory. Because of

Marine deployment and assignment cycles, with the exception of some returning leaders they are also generally inexperienced and about to learn very hard first lessons. The difference between training and a place where everyone is trying to kill you defies understanding until you've seen a man turned inside out. As a result, new units are often either too tentative or too aggressive. Sometimes they are both.

I had come to Sangin aggressive and stayed that way, but I recognized over time when that approach was counterproductive. Zak helped me understand when to pull back from confrontation. Not in a fight; that was my call alone. But counterinsurgency is a war for opinions, of nuance, of not making gross mistakes, even if you don't do anything wholly right. There were times in the villages when I knew I was speaking to someone who may well have shot at me or my Marines within hours of us meeting. Zak helped me plan my role in those interactions so that, in the confusing environment of counterinsurgency, even if we did not win, we also did not lose. Now my job was to try to help the incoming Marines learn as much as possible in a few days so they could improve upon the conditions we handed them through the blood sacrifice of the best among us.

First Platoon was sadder and wiser for our surplus of hard-won experience. We knew our craft well: the difference between incoming and outgoing fire; how to respond to Taliban attacks by gaining enough fire superiority to close with and kill them; whether a weapon's report was close enough to matter. Most importantly for Sangin, we knew where and how to move in the most IED-infested area of Afghanistan and how to get a casualty out as fast as possible.

But we were also exhausted by months of a violence level incomprehensible to most of the Marines arriving with the new battalion. Our bodies were stripped down to the bare minimum of muscle and bone required to sustain combat operations. We were emaciated from poor nutrition and little sleep, a fact made incredibly clear standing next to Marines who still had gym muscles. We were jumpy, quick to evaluate the danger of anything, from the landscape to orders from a company headquarters that had largely lost our confidence. Worse still, my

Marines were beginning to envision a return home as something more than a remote possibility. That could distract them, dangerously.

Combat Marines must exist in the moment. Thinking too much about tomorrow is as deadly as the inexperience of the newly arrived unit. The moment found us on our final patrol of the RIP/TOA, at Patrol Base Vegas, a shitty squad outpost bordered on one side by a 500-by-500-meter floodplain we called the Golf Course because nothing grew upon it but short grass. Even through March, we continued to operate as if all we needed to do was show the villages that we offered the best alternative and they would say, "Oh, we're all in with you." That was nonsense, but we could take a Hippocratic approach and first try to do no harm. Zak was a key to that. Near PB Vegas was a village called Kodezay. First Platoon had developed friendly relations with the people there relative to most of our area of operations, largely through Zak's efforts.

Zak was as exceptional in the cultural realm as he had been in a firefight. He helped me with cultural sensitivity, but he was also a human bullshit detector and helped me at least know when I was being lied to, even if I could not do anything about it. That insight helped us tailor our actions to support the villages that were at least partially inclined to reject the Taliban, even if they were not particularly inclined to accept us.

Zak translated what we were selling into concepts both palatable and desirable to the locals, and ultimately, partially through his feedback, we recognized we were making a difference in Kodezay. We had resources like food, jackets, programs for families, and medical care and they hedged their bets with us in the short term. The people asked, "Which side can give me the best opportunity to feed and secure my family right now?" I think the locals saw that by January, it was the Marines. Markets reopened, we began to see a lot more women outside of houses than at the beginning of the deployment, and girls began to go to school.

I don't think they were betting on us for the long term. I don't think at any point people in Sangin were ready to have a Western democracy, or were even interested in Western democracy. But they were interested in Maslow's hierarchy of basic needs, things like security, food, and shelter, and we were helping improve those things once we had the security situation under control.

As we finally had time to make some headway in the villages, with the slowdown in operational tempo, we also had time to mentor the Afghan National Army. Increasingly, direction came down to emphasizing a need to partner with the ANA and build a foundation from which to transition operational control to them. Communicating with the ANA became as important, if not more so, than communicating with the people in the villages.

The problem was that some of the ANA were lazy. Zak became my agent, trying to convince them to do things I wanted them to do. Those things typically ran counter to the nothing that they preferred. I had started the deployment ready to kiss babies and shake hands but after ninety days of combat, I was less charitable than I might have been in telling the ANA platoon commander, "Get your guys up, get them here at this time, and do this kind of stuff." Other interpreters required sentence-by-sentence instructions because they did not understand the bigger picture and purpose. Because we were friends and had already had conversations about what I wanted and needed from the ANA, Zak saved me from myself by massaging how I was received. It gave him an ability to explain and sell my message when he was visiting with the ANA as a fellow Afghan, rather than as an emissary delivering a message from the American. That constant message reinforcement helped build a sense that I respected the ANA rather than used Zak as a go-between and they began to join us on missions more readily and enthusiastically. That was a good hand to give my replacement for him to play.

WE ALMOST ALWAYS started patrols out of PB Vegas by going through Kodezay, since the Taliban were less likely to put an IED there. They had long since learned they were better off not killing the village kids. I explained the facts of life in the moment to the lieutenant replacing me as we discussed the final familiarization patrol we would accompany them on. Most of First Platoon had already headed to Camp Leatherneck to begin the movement back to Camp Pendleton. But Sergeant Decker, Zak, my machine gun squad leader, Sergeant Nikirk, and I would serve as tour guides for a patrol otherwise composed of newly arrived Marines. The lieutenant leading the platoon replacing us was a brave, intelligent, and talented officer. But I could tell that not everything I said was getting through.

"Ninety-eight percent of the world's opium poppies grow in Afghanistan," I told him. "Helmand Province, specifically Sangin district, is the heart of the drug market that funds the Taliban."

He nodded. Training had already armed him with this fact.

"Between September and December 2010, we were in a firefight on *every* patrol. *Every. One.* For one hundred days."

Again he nodded, face impassive.

I knew what he was thinking and feeling. I had been there seven months before, armed with training-based understanding rather than understanding born of experience. Every infantry officer who truly has the calling wants direct combat. After all, our stated mission is to locate, close with, and destroy the enemy. But the truth is that it's all academic until combat is a present reality. Then you start thinking hard about the implications of everything you thought you wanted. You start feeling things that just can't be fully understood until the possibility of violent death is truly manifest.

"Before January, we found hundreds of IEDs. But, man, look, back in January the Taliban told the farmers they were behind on poppy production. The poppy farmers said they could not farm for fear of getting killed in the crossfire between us and the Taliban. So the Taliban signed a fake-ass treaty, saying they were sick of fighting, and they just wanted to join the government. They said they would clean up the IEDs, turn in their weapons, and farm."

I had the lieutenant's attention. IEDs, and their effects, are the common thread of the Global War on Terror. A reduction in their use meant fewer potential casualties for his Marines. Zak interrupted us. Now some children from Kodezay had arrived at PB Vegas to tell Zak that while we had been patrolling near the adjacent river the day before, the Taliban took advantage of our absence to spend a day digging in fifteen to twenty IEDs all over the Golf Course. That was the perfect segue for me.

"So, the Taliban and the Afghan government signed the treaty. All of us here on the ground knew it was bullshit. The terms meant we stayed on our base for several days to allow these assholes to supposedly clear the IEDs that they had laid before without us shooting them. What they actually did was turn in three rusty antique rifles, like this treaty required, then used the time to reseed the area with two to three times what they laid previously. Then they told the farmers to get back out there and farm poppy to make them money to fight us with. So when we went back out in January, there were no more firefights but way more IEDs. Which brings us to today."

My replacement lifted his eyebrows and exhaled through pursed lips. IEDs are frankly terrifying. They are not what any of us signed up to fight. As long as a Marine has someone to shoot at, he is generally okay. Marines want to fight an enemy that wants to fight them. An IED kills without recourse beyond a medical evacuation and a hope that your legs are only gone below the knee and your dick is still there. I could see his thoughts churning. Today was his patrol. In Marine lingo, he was in the left seat, the driver's seat. I was in the right seat, as a passenger and tour guide to a pastoral paradise where poppies and IEDs were both planted in abundance. And he wanted to go through the Golf Course.

I shook my head and said, "We shouldn't go through that field, man."

He looked at me and I saw the certainty the Corps trains in its leaders. "I don't want to set a pattern of always going through the village," he said.

"Dude, we always go through the village because the assholes don't put IEDs in the village."

"Not today. We're going through the Golf Course."

There was not much to do but say "Okay" and get my guys ready for the patrol.

A Marine leader is expected to be where he or she can best control the unit and affect events at the point of friction. When accompanying one of my squads, I usually patrolled in the first third of the patrol, usually as the third or fourth man. The patrol was a bit larger than normal since we were augmenting the newly arrived platoon. Thus there were five Marines across the Golf Course when Sergeant Nikirk stepped onto the field and disappeared in a cloud of smoke and mud, accompanied by a loud "POP!"

When you see one of your Marines injured or killed by an IED, it is an impotent feeling. Your enemy typically disappears long before they inflict actual damage upon you and your Marines. Unless they reveal themselves by firing at you, there is nothing to fight but time as you try to stabilize a critically injured young man, convincing him to hold on through evacuation to the next level of care, and hope that you did not miss anything vital as you evaluated his injuries. I ran to Nikirk, with Zak on my heels as always. There would be no need for him to interpret. He had simply become one of us over the months we spent together and he was preparing to help me attempt to hold someone together as a Navy corpsman rendered lifesaving first aid.

But as the smoke cleared and the mud and dust settled from the air, I arrived at Nikirk's side to see he was standing next to a partially ruptured, forty-pound jug of homemade explosive, much of it now dusted across his face. He was alive, perceptibly shaking from the experience of a low-order, partial detonation of ammonium nitrate homemade explosive (HME) intended to sympathetically detonate a 105 mm howitzer shell right under his feet. The combination should have left Nikirk a pink vapor drifting in the air with the lingering smell of ammonia. But because the area had recently experienced a lot of rain and Taliban quality control was low, the IED pushed him aside, plastered his face in the ammonium nitrate and aluminum used to make the explosive, and exposed the artillery shell that would have left him nothing but a memory.

Since we had barely left friendly lines, and it had become clear that Nikirk was largely unharmed, I called for Explosive Ordnance Disposal (EOD) to come out and destroy the IED. After an hour, they arrived and confirmed that the IED should have meant the end of more than one of us. They also noted that the IED was triggered by a tripwire, the first we had seen of such. Typically, IEDs in Sangin were triggered by the victim's weight pushing down a pressure plate, which completed an ignition circuit.

Sergeant Nikirk and Sergeant Decker came to me and asked to return to base with EOD. I sent Nikirk back but told Decker he would have to remain. We needed his experience if things continued to go downhill. Sergeant Decker looked at me, reminded me he had a son whom he had never met, and said, "This is fucking bullshit, sir. These guys are going to get us killed." Decker had never needed more than a direction and distance to, and description of, the thing I wanted attacked and destroyed. He was both courageous and cautious, a force of nature in combat. Now, for the first time in his life, Sergeant Decker was ready to pack it in on a combat operation. I told Decker we were continuing on, then turned my attention to my incoming lieutenant counterpart and asked him his plan.

"We're going to continue to push across the field."

Aggression aside, I was stunned.

"The only way we are going across that field," I said, "is if you have the combat engineer back-clear from his position to us, then reclear it, since he had obviously missed at least one of the IEDs."

Combat engineers accompanied us on most patrols and carried a metal-detecting sweeper intended to find IEDs and land mines. The incoming platoon commander gave the order.

We had been stationary for more than an hour as we dealt with Nikirk's IED detonation, then EOD's arrival and departure with Nikirk. That was way too long and now the Taliban certainly knew exactly where we were as the combat engineer began the exhaustive process of sweeping back across the flat, open expanse of the Golf Course from five hundred meters away. He was halfway across the field, coming back to us, his sweeper ticking back and forth like a metronome, when the second IED blew. This time it was a complete detonation.

The combat engineer disappeared in the fire and mud and smoke. Before the debris had stopped falling, the Taliban unloaded on us with rifles, rocket-propelled grenades, and medium machine gun fire. They knew someone had to go get the combat engineer. They knew there were a shit-ton more IEDs in the field. They wanted us moving around to hit them.

With the incoming fire everyone hit the deck. You could tell the difference between new guys and 3/5 Marines. The new guys were face-down in the dirt or neck-deep in an irrigation canal, rifles firing in no particular direction. Those of us on our last patrol were up on an arm, scanning for muzzle flashes and smoke. Painful experience told us we had to determine the source of the fire and put our own on them before they could hit one of us.

I was furious. This was our last patrol and after seven months, stupidity was going to kill us. I looked at Zak. All I could say was, "Son of a bitch! Can you believe this shit!?" He just shook his head in disbelief and said, "Lieutenant Tom, this is crazy." Of course, Zak wasn't leaving with us. He would stay here.

As I continued to scan for the enemy, out of the corner of my eye I saw Sergeant Decker run onto the field toward the wounded combat engineer. The man who had asked me to let him return to PB Vegas, the one who reminded me he had never met the child he'd named Maximum Danger Decker, was running into the midst of an uncleared field planted with double digits' worth of IEDs to retrieve a wounded Marine he didn't know. I thought about the fact that it was our last patrol, that I had, only moments before, denied Decker a chance to return to safety, and that I now assumed I would soon be living with the fact that I denied a dead man a chance to meet his child. I screamed, "DECKER!!!" He kept moving into the field. I screamed again, "SERGEANT DECKER! STOP!"

He looked back. I had no children.

"Come back here! You're not going, I am!" I yelled.

Decker started moving back to me. I looked at Zak and asked, "Are you ready to go?"

Of course he was. Zak's eyes were open a bit wider than normal, but he was always ready to go. Even when, as was the case now, there was no reason for an interpreter. I just needed an extra set of hands to save an American life. I looked at the canal we had to cross. It was frothing from the bullets striking the water's surface, as if a tropical rainstorm had set down in Sangin, but only on the Golf Course. I looked at the rest of the patrol, spread out in single file and hugging the earth with their dicks in the dirt, looking at nothing, just spraying bullets everywhere in a death blossom.

"Fuck!"

Zak and I were the only people up. We made convenient targets for the Taliban machine gunners as we ran up the column, me screaming, "WHERE IS THE CORPSMAN!?!?" The guy tasked with providing lifesaving care should have been up and moving already.

I saw a hand go up, inches above the dirt, his face pressed into the mud. I grabbed the drag strap on the back of his body armor, yelling, "Follow me!"

Time slowed down. I was moving forward, dragging the corpsman into the field toward the combat engineer, the extent of whose wounds I still did not know. Zak pushed him from behind as we all winced against the incoming hail of steel. I was thinking about what we needed to do. Simultaneously I thought, "My last fucking day! My last fucking day! Best-case scenario, I am not leaving Sangin with my legs. Worst-case scenario, I'm gonna be turned into pink mist by a 105 shell."

There are often absurdity and seriousness in equal amount during combat. As we ran into the field, knowing that the first IED had been initiated by a tripwire, something we had never seen in Sangin, I was running like a football player doing high knee drills, trying to avoid additional tripwires. Every time my right foot struck the ground, I yelled, "Motherfucker!" like some absurdist running cadence. For 250 meters it was, left foot, "Motherfucker!" left foot, "Motherfucker!" left foot, "Motherfucker!"

We reached the casualty and I threw the corpsman at him so he could begin to do his job. Rain and the weight of the mud had again been our friend. The blast had thrown the combat engineer through the air but the weight of the mud had tamped down the explosion. His major injury was an arm bone sticking out through his flesh, a relatively benign result.

I had been carrying an M32 grenade launcher for two months simply because I wanted to use it in a firefight. Imagine the world's most powerful revolver as a six-shot, rotary-magazine, 40 mm weapon. Now I had my chance. It seemed like the thing to do since I still expected to die recrossing the field. As the corpsman worked on the wounded Marine, I started slinging grenades.

With 40 mm explosions not to their liking, Taliban fire slackened to an acceptable level. The corpsman pronounced the combat engineer ready to move and we headed back across the field without hitting any more of the IEDs we knew were there. I got back to the incoming lieutenant and hissed, "This patrol is over!"

"I guess I should have maybe listened to you on the route." No shit.

My grandmother died of emphysema. I was a good kid. Never drunk in high school, never in trouble. I'd never had a cigarette in my life. We got back to PB Vegas and my first words were "Who has a cigarette?"

I stood and smoked my first cigarette, a Camel Blue, with Zak and it was so, so good. I never coughed once.

LOOKING AT ZAK, as I departed FOB Inkerman, was like leaving a blood relative and knowing it was likely forever. We had experienced the entirety of the human condition together, had fought together, lost friends together, and laughed at things never considered by the main of humanity. It's hard to put words to the experience of a combat deployment, especially ones in which accomplishments and benchmarks seemed like sand castles made by excited kids, one moment standing proud, the next crumbling before an inexorable tide. In the absence of clear and sustainable results, the Corps generally falls back on the numbers we call "metrics" as a means of accounting for the outcome of operations. The "metrics" sanitize First Platoon's seven months of lived reality through a sterile reporting of military events that ignores the radically altered life trajectories of entire families. Between September 2010 and April 2011, I, Zak, and First Platoon lived bloody lifetimes behind the numbers.

First Platoon executed 167 squad-sized patrols and another twenty platoon-sized operations. Three Marines were killed in action under my command. Sixteen, more than a third of us, were wounded by gunfire and explosions. Sometimes that meant a few days of rest; sometimes it meant traumatic amputations and a lifetime of pain. Along the way, we found 102 improvised explosive devices, sometimes by triggering them. We found eighteen weapons caches filled with rocket-propelled grenade launchers, rockets, rifles, vast quantities of ammunition, and thousands of pounds of homemade explosive and other IED components. In months of daily firefights, we killed seventy-five Taliban, wounded nine, and captured another twenty. Zak put himself and his family at risk of assassination forever, simply by trying to make a better life for the people of Afghanistan. The metrics totaled to equal the fact that I was ready to go home by April 2011 and, even if I wasn't, it was time. Orders are orders. But I hated to leave a friend to a fate over which I had no influence.

On April 6, 2011, I walked from the helicopter pad at FOB Inkerman to Zak's room. He didn't have a door, just a blanket. I slapped my palm against the mud wall outside his door and called for him, "Zak! Hey, Zak!"

He pulled the blanket aside and gave his customary grin. "Lieutenant Tom. I am glad you are not gone to America yet."

I did not know what to say. What do you say to the man who stood ready to put himself between you and a bullet? How do you say goodbye to someone you love and whom you also know you will likely never see again? How does a warrior say good-bye to a war and the brother he is leaving with it? Every word that came to my mouth seemed inadequate.

"Hey, Zak, I wanted to say thank you. You have been amazing. I will be there for you if you need me someday. These new guys will treat you right. But either way, I just wanted to thank you, brother. Here's all my points of contact." I handed him a page torn from my notebook, my phone number and email written upon it.

"It's all good, Lieutenant Tom. Thank you. You and my brothers in First Platoon, you have done so much for my country. You are my friends. You fought for my country to have a better future. I am proud of you."

I felt like there were a million things to say, but I had to get to the helicopter landing zone for my ride out with Nikirk and Decker. My gear was already there. We would fly to Camp Leatherneck to meet the rest of the platoon who were already there. Elements of the battalion had already flown to Manas, Kyrgyzstan, and would soon be drinking cold beer in Southern California. We were both silent a moment before Zak said, "You better go before you miss the airplane. I will walk down with you."

We walked together in silence. There was no need to speak of the moments we had already shared. I was unable to speak about the different futures that awaited us. We reached the landing zone, where an MV-22 Osprey was already landing to take us aboard and fly before the Taliban could fire rockets and mortars at it. The rotor noise was growing as the Osprey settled onto the landing zone. I just needed to pick up my gear and board the bird. I was ready to get home and fix my failures with Andrea, but part of me wanted to stay in Helmand. Part of me always will.

Before I could, Zak had already shouldered my stuffed ruck sack and started walking to the plane. In Helmand's blistering heat, the Osprey's prop wash and engine exhaust were even hotter than normal. Under the tail, before turning to walk up the ramp, I reached out and pulled Zak close for a hug. I did not trust myself to speak for a moment.

Still gripping his shoulders, I looked down into his eyes, then leaned in to shout in his ear, "Anything you need, brother. Anything."

He nodded and smiled. I squeezed his shoulder and fist-bumped him before walking up the ramp, taking a seat on the webbing benches, and turning to see him walking away to the edge of the landing zone.

The Osprey rocked and vibrated as we began to break free of Helmand's grip. The ungainly plane/helicopter combination gained power and altitude, turning toward Leatherneck as the Osprey's nacelles rotated forward for flight. On the landing zone, I watched Zak shrink as we flew higher and then disappeared in the dust and shimmering heat of FOB Inkerman. I turned my face forward to take advantage of the slipstream through the cabin to dry the rivulets of sweat coming from under my helmet and thought of what might become of my friend. Perhaps I would be back for another tour, perhaps not. But whatever my distant future held, in the next few days, I would be in America and Zak would still be at war in Helmand. There was no redeployment for him.

My Marines were covered in tattoos. Death Before Dishonor. USMC. *Semper Fidelis*. Those are fine ideals to live by. They come cheap in tattoo joints all over Oceanside, California. But for Zak, life in Afghanistan really was going to be victory or death. It would be another decade before we realized how true that was.

PART II

THE ESCAPE

11

New Beginning

ZAK

I DIDN'T CARE about the sand that flew in my eyes as I watched the huge, gray Marine Osprey lift off with my brother, Lieutenant Tom Schueman, as he departed FOB Inkerman in April 2011. He was sitting in the rear door, waving good-bye to me as the aircraft rose into the air. I wanted to remember the moment and all the ones before it. We had fought too hard for them to let any be forgotten. To live the way we had each day, not knowing if we would live or die or be mutilated, created a love that made it hard to accept that I would probably never see him again.

The nose of the aircraft dipped toward the earth as it lumbered forward, then gained lift and rose into the sky. I thought back to the time before I came to Sangin, back to conversations with my father. I realized that with First Platoon I had done the things I came to do. We fought the Taliban. We protected the people from IEDs and attacks. We tried to make a better future for Afghanistan. We learned lessons that kept us alive. Sometimes we were just lucky. Sometimes we lost our friends and brothers along the way.

I lifted my arm and waved at Tom. He waved back. I watched the helicopter fly until he was out of sight. He waved the whole time. First Platoon was gone. Tom was gone. I had been beside him for seven months and now I was alone in Sangin. I missed my Marine family and I missed my real family. Walking back to my room, I looked out at the brown hills of Sangin from the walls of FOB Inkerman. I thought of Kunar's mountains. I knew my family still needed money, but I was ready to go home. I called my father for his wisdom.

"Father, my team is gone and I am tired. But I can stay with these new Marines. I know we need the money."

"Zainullah, there are jobs here also. It is too dangerous there in Sangin. Come home, son."

"The money is good here."

"Son, you cannot spend money if you are dead. If the Taliban has Sangin, we will have Kunar. Come and help at home. You will find a job in time. We want you safe. Your mother cries for you."

"All right, I will call the Mission Essential Personnel contact at FOB Jackson and resign." I went to my room to pack. I was going home.

MEP SAID THEY would give me work in Kunar if something came open. In the meantime, they flew me to Camp Leatherneck and on to Kabul on American aircraft. But this time there was no bus convoy. Once I was in Kabul, I had to hire a taxi. The threat of Taliban checkpoints had increased. After most of a year fighting them in Helmand's flat dirt, cornfields, and desert wadis, I was anxious riding through the rocky hills and river bottoms of Laghman Province and through Jalalabad. The area between Kabul and Kunar feels less like the surface of the moon than Helmand, but the dusty hills and rocks of Laghman look enough like the west side of the Helmand River to stir the instincts that kept me alive in Sangin for seven months. The countryside grew more comforting the closer we got to Kunar and the peaks of the Hindu Kush, but I did not feel calm until I reached our home in Asadabad.

For months, I rested and worked at home. It was good to be back with my family, but it was hard to leave Sangin behind. Just like my leave period before, I still heard shooting in my head when I closed my eyes. I still watched where I put my feet. But there was no one I could talk to about that, so I put it away as much as I could. It came back when I slept. But I was with family when I was awake and that made me happy.

Sometimes you have to leave the things you have to see them clearly. Waking up in the semidarkness of the bedroom I had known all my life, I felt safe. There would be no firefight in our yard today. That seemed like enough after what I had seen in Helmand, a place full of people with almost nothing, wondering if they would lose that, too; a place where violent, dangerous men had power over terrified people who just wanted to live. Unfortunately, I also knew that the sense of peace I felt at home in Kunar was artificial. I had seen the Taliban face-to-face in Sangin. I knew what they could do and that they were only a few miles away, in valleys like the Korengal and Watapur. I knew they waited just over the border in Pakistan.

Our family was blessed by Allah to have the things we had. But I knew from talking to the Marines that I wanted more of what they had for my country, especially freedom of information. I could never again allow anyone to force me or my family to live in ignorance of the world outside Afghanistan. I could never allow anyone to threaten me for wanting *to know*.

My time in Sangin increased my desire to go to university. I knew the fighting in Helmand slowed down only because the Taliban chose to slow it to continue their war, their way. I knew they would bring that war to Kunar if they could. I did not see a more effective way to defeat the Taliban than building education and knowledge. I thought of my teacher, coming back to Asadabad from Pakistan when the Taliban ran from the Americans. I began to think I might like to be a teacher someday.

My oldest brother, Asadullah, had married and brought his wife to live with us as well. My father was planning to expand the house and build another bathroom for them. He was working at a radio station in Asadabad, managing the programming for information the government was passing to the people. Sometimes I helped him out, but I did not have an actual job. With so many people to feed, soon my parents needed me to work again. In December 2011, I heard there was work for interpreters at FOB Wright. I enjoyed developing my English skills and meeting Americans, so translation work was good for me. The more English I could learn and perfect, the better I could share it. Translation was also the best-paying work available. It had been a year and a half since I had been to FOB Wright and begun my journey to Sangin. Going back offered the best chance for my family's future.

When I got there, I was not the only one trying to get a job. There were about fifteen of us. We made small talk while we waited. Some of them were like me and had already worked with the Americans. Some were younger people, looking for their first job as a translator. As I had done more than a year before, I stood with my hopes and waited for a chance to work. Only now there was a difference—now I was the nervous one in the line. I knew what a bomb could do to a man.

The Taliban had made gains over the year I had been gone. In June 2011, not long after I left Sangin, U.S. president Barack Obama announced a plan to withdraw 33,000 troops by the summer of 2012. The first 10,000 would be gone by the end of 2011. It was not a secret that the United States had begun peace talks with the Taliban. First Platoon had been strong. We fought hard in Sangin. Now it seemed that there would be a deal cut with the same people who ran into the mountains to escape the American and international forces ten years before. I remembered my teacher all those years earlier who had excitedly told us about Taliban jets attacking buildings in New York and I wondered if he would be back in my village.

Standing there in my hometown and waiting for the contact at FOB Wright, I watched men riding two at a time on motorcycles, a common sight, there and in Sangin. My brain looked for signs that they might be emplacing IEDs or planning an ambush. Fortunately, nothing happened before our contact came. He let us into FOB Wright for another test of our abilities to read and answer questions about what we read in English.

When the testing was complete, an American came to speak to a smaller group of us. The gray hair under his baseball hat and his red beard with gray streaks made him look older than he really was. He did not give his name, but he seemed like a nice man. He was not in the military. He said he was from the United States government and was in charge of support for their operations. He called us into a small office individually.

When it was my time, I sat down with him to talk about the job. I gave him my MEP employment records and a recommendation letter I had received from the Kilo Company commander. He read all of it and said, "You look like the perfect man for this job. I need a translator to work outside the camp at first. I have thirty laborers building brick walls. I need someone to explain their daily tasks and supervise them in the execution. I also need someone to explain what's going on in the crew and translate for me when I need to get information to them."

I knew I could do that and said, "Sir, I worked in Nangalam for seven months building the district center there. I can help you."

He looked pleased. "Then the job is yours. Once we are done with the job outside the camp, if it works out, I will bring you on inside the camp full-time."

The work was fine that first two months, but once I came inside the camp, it was really a good job. The Americans brought good employment for many people—Afghan carpenters, welders, drivers, mechanics, cleaners, and cooks were all happy to have the work. That made my job easier. Working for the U.S. government paid better than working for MEP.

I lived at home with my family and walked to FOB Wright each morning to provide translation between the Americans and all of the Afghan people on the base. I worked in a clean, air-conditioned office and was the assistant to the Afghan translator who worked for the American director of facility services. He handled the paperwork and payroll in the office as well as all the translation related to negotiation with service providers. I assigned and explained the jobs to be done to the workers every morning and evening. There were no patrols, no IEDs, and no one shooting at me. I started to think that even if what

I'd lived in Sangin was not what I envisioned for my country, Sangin was a long way from Asadabad and there could be a good life in Kunar for people there who wanted it.

I understood my country. I did not think that we would become the United States and I did not want us to. I am an Afghan. I am proud of our culture. It has survived thousands of years and many invaders. I truly wanted a life in my home province. I wanted to have a family in the place I loved. I wanted an Afghanistan where free, honest government processes supported people's abilities to lead their lives happily, with honest leaders who cared about the Afghan people elected in free and fair elections. I wanted schools that taught things people need to know to work in the modern world or to make the most of traditional Afghan trades and crafts. I wanted a nation where the overwhelming spirit was not "get what you can, when you can." I went to Helmand with hope that if we could make a difference there, we could certainly make other parts of the nation, like Kunar, a beautiful place to make a life.

Though I was hopeful, I was realistic. As well as things were going in my job, like all Afghans, I kept one eye on the news. I had been in Asadabad in 2002 when the Americans began coming in large numbers. I was in Sangin when the Americans "surged" there. Now they were leaving in large numbers. Twenty-three thousand American service members would depart by September 2012. Less than a year later, in June 2013, the government of the Islamic Republic of Afghanistan took over security for the entire country from the international forces. The U.S. government people I was working for were leaving. That meant my job was leaving, too.

In early 2014, with the Americans departing, I was terminated from my job at FOB Wright because there was no more work for me to do. Most people I knew in Asadabad were employed with the U.S. forces or in Afghan projects paid for with U.S. money. Suddenly we had nothing. There were no jobs. It was time to take new steps if I wanted a professional job: I needed more education. My brother Asadullah, who was working three jobs, gave me money to begin studying business administration after work each day at the Tanweer Institute of Higher

Education, in Asadabad. I did not have a job and, for the first time since 2010, I did not have a plan except for finishing my degree.

By the end of 2014, all American troops were supposed to be out of Afghanistan except for trainers helping the Afghan National Security Forces (ANSF) secure the nation. That meant that Americans would no longer be leading the fight against the Taliban, Al Qaeda, and the Haqqani Network, to name the main enemies. It would be up to the Afghan National Security Forces. When I thought of the mix of ANA at FOB Inkerman, I did not think much of our chances. There were many brave ANA soldiers. But there were enough who wanted to smoke hashish and sleep instead of patrol that the brave ones had difficulty.

It still seemed impossible in 2014 that the ANSF could truly fail to defend the nation against the Taliban resurgence after the Americans left. There were 320,000 men and women in the ANSF. But without the U.S. presence, it seemed likely that the Taliban would return in some form in the rural areas. I felt sure the ANSF would hold the major cities and keep the Taliban back from places like Asadabad. But I knew they would still come in the night and that I and my family would be in danger because of them.

It was always dangerous for Afghans who worked to support the Americans and the Afghan government. The Talibs were already leaving threatening letters for people like me. "Night letters" stamped with the official symbols of the Taliban and posted in public places threatened death to people helping "the occupiers." They usually listed some specific names and said they had to turn themselves in to the Taliban shadow courts to be judged and punished. I knew that meant death. If I did not turn myself in, they would try to kill me anyway, so it didn't matter. I had not yet seen a posted letter with my name on it. That made it an easy choice to ignore their directives, but I was scared. Without our American friends to stop them, the Talibs would have no restraint. In Sangin, I was with Marines I knew would protect me. Now, at home in Kunar, I had no one.

I began thinking about leaving Afghanistan, but since I was jobless, there was no money to move. Pakistan was only eight miles over the mountains, but the Taliban were there and could easily target us. The government of Pakistan wanted nothing to do with people like me who had fought for a safe, strong, and secure Afghanistan. They were supporting the Taliban. I was out of choices. I did what I always do. I went home and asked my father for his opinion.

"Father, I don't know what to do. I don't want to go to Kabul, but there are no jobs for me here."

"Zainullah, we survived the Taliban. Now we are in a better country. Things will work out."

"Father, I have been working for six years. I want to finish at the university. I want to make a better life for our family. I have wanted this since I was eighteen. There was no money before, but now I can go and make a better chance for myself and our family as I have made a better chance for Afghanistan."

"My son, you can certainly go to school and live here, but it is time to consider your own future as a man and the head of your own house. I have been to see my friend Mr. Hamid from my time in Dr. Najib's government." Hamid lived in another district of Kunar, about six miles from our home. "I went with your uncles and your mother. We have seen his daughter. She will be a good wife for you. He has said for us to come back to discuss the bride price and the marriage ceremony cost and he will make a decision and let us know if it is possible."

Now I had even more to think about. I had no job. I had no house. But my family had found me a wife. In places where marriages are not arranged, that may seem strange. But I was twenty-four and well past eligible. I was only surprised it took so long. I was excited to marry and begin to live as a grown man. I was also nervous and hoped it would be a good marriage.

My father and mother and uncles returned from Mr. Hamid's house to tell me I would be married to his daughter. I had only a little money for jewelry for the wedding, but I went and got a ring, a necklace, and earrings. Pashtun culture meant my mother would give my new wife some of her jewelry as well. We set a date and sent out invitations to relatives and friends. I remembered my brothers in First Platoon of Kilo

Company and wished they could be there. I wished Lieutenant Schueman could be there to stand with my older brother at my wedding. I may not have been alive to get married without them.

When the day came in April 2014, I woke early, well before my prayers. I was too nervous to sleep. I slipped from my bed to avoid waking the three brothers who slept in the room with me. I went outside into the light of a full moon. In Sangin I had stared up at a full moon and wondered how I could live such a different life from people staring at the same amazing sight. It seemed impossible that my parents in Kunar could be looking at the same moon as I was in Sangin, while I was living a life they thankfully knew nothing about. It was even more difficult to think that a Taliban fighter could look at the moon the same way as I did and still want to place explosive in places they knew would mean another young man would lose his leg or his life. I tried to put that out of my mind and focus on the fact that I would be married by the next moonrise.

Even though I grew up in the very traditional culture of Kunar, I was still uneasy that I had not yet met my wife. I was concerned that she might not like me. Maybe I would not like her. It didn't matter; my father and Hamid had decided. In big cities like Kabul, a bride and groom can sometimes meet each other and talk. They might talk to one another on the telephone. In Kunar, that does not happen. By tradition, I would not meet her until the ceremony later that day. I sat until the sun began rising, then snuck back to my bed so my brothers could not tease me about being nervous.

Guests arrived at our home at five in the afternoon. My bride was in another room in our house. I had still not seen her. Soon the mullah arrived to discuss the *Nikah*, the marriage contract, with me and my bride's father. The *Nikah* is negotiated in private. Only I, Mr. Hamid, and the mullah went into my parents' bedroom to discuss it. There I finally learned my bride's name, Deewa.

The mullah read the terms of the marriage contract, then three times he asked me whether I accepted the terms in it. He then went to the room where Deewa and my father waited. He explained the contract to her and asked three times whether she accepted the terms. When she said yes, we were married and could finally meet.

I stepped out of my parents' bedroom onto the terrace as Deewa came out of my sisters' room. There was music playing and our relatives were clapping and cheering for us. I looked at her and felt good. She was beautiful to me. I was happy to be her husband. As we came together, our mothers covered us with a shawl. The mullah gave us a verse from the Holy Koran to read together. I looked at her and smiled. She smiled back as we joined in reading: "And one of His signs is that He created for you spouses from among yourselves so that you may find comfort in them. And He has placed between you compassion and mercy. Surely in this are signs for people who reflect."

As we finished, our mothers returned to hold a mirror in front of us. With my wife and I looking back at ourselves for the first time, I saw myself as a married man, reflected back with the woman Allah had sent for me to love. It felt good. I was proud.

As at my graduation six years before, we had a huge party. Afghan weddings are a reason to eat. My father bought and slaughtered a cow for the feast. There were huge dishes with rice and beef, vegetables, and sweets. Like my graduation, the celebration was separate for men and women. As the evening progressed, Deewa and all the women guests stayed at our house to dance and talk. I went with all the men to the village guesthouse, a common building for people to use. As we walked, men from my village teased me about my marriage night. I was embarrassed, but I laughed with them. It was part of being a Pashtun man. In both locations, the party continued till late, with men and women again dancing the *Attan* separately. We spun and laughed until it was time for me to go and pick up my bride and start our lives together.

I WAS HAPPY to be married to Deewa. I looked forward to building our family, but I still had no job. That fact made her parents nervous about my ability to provide for a family. It made me nervous, too. Without a job, I did not have the ability to buy or rent a house. Deewa and I moved into my parents' home. This is common in Afghanistan. When sons take a wife, our families just build another room onto the son's father's house. We had already done so for my brother Asadullah and his wife, even though he had three jobs: one at a radio station in Asadabad; another working for the Organization for Social, Cultural Awareness and Rehabilitation, teaching nonviolence; and another as a contractor teaching English in Asadabad on behalf of the U.S. embassy. It was a good thing he did have so many jobs. He and my younger brother Faizi, who worked in the Department of Agriculture, were now supporting thirteen people.

Of course, all of us contributed to the household in some way. My sisters, Deewa, and my brother's wife worked in the home with my mother, cooking and cleaning, and feeding the animals. My sisters also went to a school funded by the Afghan government. After years of women being denied any kind of education, my parents wanted them to get as much formal education as possible so they could be successful in the new Afghanistan. All of them eventually graduated from high school but, like me, there was no money for a private university after that. My youngest sister did eventually attend a government-funded madrassah for further Islamic education. It was not a university, but the Ministry of Education paid a woman in our village to open her home and teach the Koran to women and girls in our area. Not every woman went. Even thirteen years after 9/11, provinces like Kunar held on to tradition. In Afghanistan, most families don't want to send their sisters and daughters to work outside the home. Men are expected to work outside the home and provide everything in that regard. That made my unemployment a problem. I was expected to provide for my wife, my mother, and my sisters but I had to rely on my two brothers and their jobs to do so.

We were not starving, all of us had food to eat, but I was always looking for work. With almost all the Americans gone from Kunar, there was no translation work. I knew my experience managing construction and civic work could help the Afghan government continue to build the country. Continuing to serve the country and help it move forward appealed to me. With the international forces pulling back and the Afghan government more fully in charge, I felt it was even more important. I answered every announcement for every Afghan government job I thought I was qualified for and some I was not. That hiring process was where Afghanistan could never save itself from its own corruption.

An Afghan government job announcement meant an examination and interview for any candidates applying. It was always an obstacle to anyone without the money to pay bribes to get hired. Bribery is a way of life in Afghanistan. Generally, the first candidate to get to an interview with enough cash for the bribe gets hired. It's even faster if there are family or tribal affiliations on top of cash. Sometimes that means a score on an examination gets increased. Sometimes it means the scores for ones above it don't get recorded. Of course, the hiring announcement reports that the best candidate has been hired. It sounds funny to say, but I didn't have enough money to get a job.

It was even more frustrating after working with Americans for four years. MEP hired me and sent me to Sangin because they needed the skill that I had to help the ANA and Marines make Sangin safer for the people there. At FOB Wright, the American from the U.S. government hired me because I was perfect for the job. Now, at home, I compared it with my inability to buy myself a job with the Afghan government.

Corruption was a termite, boring through our national foundation and threatening to collapse the institutions we had built since 2001, though the problem went back decades before that. In 2010, there was a run on the bank when the Afghan people discovered that the Kabul Bank chairman and President Karzai's brother, among others close to President Karzai, were spending depositors' money to fund their lifestyles. One billion U.S. dollars went missing from the bank that handled payments to the ANSF and Afghan government. A financial system runs on faith, just like a government. Corruption destroys that. The Kabul Bank was an extreme example, but the bribery and cronyism that were normal to everyday Afghans gave the Taliban a weapon to use against the system we were trying to build.

I wanted an Afghan version of the American dream. But the Afghan people have been so abused, for so long, that many take every opportunity to prepare for future tragedy. Our tribal, familial, and village politics pitted people against one another. Things Americans take for granted were almost impossible in Afghanistan. Just paying the military was a problem. The Americans gave some of the Afghan commanders the money to then pay their men. The commanders reported one hundred soldiers. In fact, there were only eighty. The commanders pocketed the rest. We paid extra for every interaction where someone could extract a bribe. A government job was as much an opportunity to demand bribes as it was a salary.

I wondered whether the brighter future I fought for in Helmand was even possible, whether we would ever be able to defeat the corruption that was part of every level of our government. First Platoon had fought the Taliban hard enough in Sangin that the Taliban stopped fighting them directly. Now it seemed clear we Afghans might defeat ourselves with corruption, no matter how strong the ANSF became.

12

Returning

ON APRIL 20, 2011, I landed back on U.S. soil. I was twenty-five years old and had just lived a lifetime on fast-forward. My body was battered and my brain was traumatized by multiple blasts, the type of injuries sustained by so many of us in the post-9/11 wars. But I also came home from Sangin exhausted, embittered, angry, and unsure how to express any of it. There was no way to physically represent the moral injuries I had sustained, and I did not know a way to talk about them, though they were evident to anyone who knew me. I was still angry at what I felt had been Kilo Company headquarters punishing me for aggressively implementing orders I was given. I was depressed by the death and injury of my friends. I was physically wrung out from seven months of insomnia, long daily movements under load, and the accumulated stress of leading young men in a place where someone was trying to kill all of us every day. Strangely, like an addict having withdrawals, I still missed the almost daily combat of the first four months of our deployment.

Worse yet, just as an IED does invisible damage to Marines as its shock wave passes through their soft internal organs, people in my orbit were suffering from my injuries. In the week before we got back, my mom was counting the hours through sleepless nights. At work, her inability to stop crying in a bathroom stall got her sent home for the day to recover. When she met me in Southern California on return, she tried to talk to me about how I was feeling. I didn't want to talk about it. My desire not to talk never really mattered where my mom is concerned, especially after she took it upon herself to read my journal from Sangin. I was angry at the violation of my privacy.

"Thomas, I can't understand what happened over there, because you won't talk about it."

"Mom, I'm fine."

"Then please make eye contact with me when I am talking to you!"

"Would you please just let it go!?"

"Thomas, you're my son and my best friend. You have been since you were born. It's like you're not there anymore. You're just this angry person and I'm on eggshells. I'm not letting anything go."

"What do you want me to say, Mom? I said I'm fine!"

I wasn't fine. But I did not know how to not be. So, I was "fine" when anyone asked. I value order and self-discipline. I had, and have, lived my life in rejection of explosive anger and unchecked emotion. I knew I could control it if I worked hard enough. I just had to suck it up and push through.

After a deployment, there is always a postdeployment leave period, usually four weeks total, during which half the unit has a two-week block of leave each. Two weeks after we got back, I commenced my leave and went home to Chicago, where, deserved or not, I got a hero's welcome. My mom's police buddies and motorcycle riders from the Warriors' Watch Riders escorted me to TR's Pub on 111th Street. There were family and friends there. The news showed up. Two decades as a police officer had given my mom her own service family. Members of the Chicago Police Department packed the place. In the context of the hero's journey, it was "The Return," in which the hero comes home enlightened. Perhaps I was, but a lot of my enlightenment was things I wished I didn't know. I was equally burdened by unanswered questions and unexamined injuries. All I really wanted was to be alone with Andrea to try to make up for the way I treated her before I deployed.

Once I completed two weeks of postdeployment leave, I headed back to Camp Pendleton. 3/5 made me the executive officer (XO) of Lima

Company, the second in command. It was a pretty normal step following an infantry officer's first deployment, but leaving Kilo Company sucked. Worse yet, we were scheduled for another MEU rotation, this time to join the Fifteenth Marine Expeditionary Unit to float in the Western Pacific for seven months. We had a year of individual and unit training ahead of us before we officially joined the MEU for another six months of training, followed by the actual deployment in December 2012. But I had avoided three months of shipboard life with the Thirty-First MEU, and I was not excited for seven months with the Fifteenth. There was still a war in Afghanistan and I wanted to fight.

Though I was seeking more war, at the same time I was seeking peace. There was tension in the dissonance. Andrea and I had resumed email and satellite phone communication after I called and begged her to take me back at the height of the violence in Sangin. We were practiced at making up. It wasn't the first time she'd taken me back. Upon my return from Afghanistan in April 2011, she joined me in California. A few months later, in July, we went to the Orange County courthouse and made it official with a marriage in front of the justice of the peace. That was the start of the greatest year of my life. All the college boy infidelities, my unwillingness to commit, the cowardice to which I had subjected her, were gone. In the Bible, in the book of John, Jesus applied mud to a blind man's eyes and his vision was restored. My fears and selfishness had blinded me for years. Combat in Sangin alongside Zak and my Marines, and my resultant conviction that if I somehow survived then I owed Andrea and the world a better man, had been the figurative mud on my eyes. In marrying Andrea, I felt as if my vision had been restored.

I owned a three-bedroom condominium in San Clemente. Andrea and I had the master bedroom. I rented the two extra rooms out to a lieutenant with whom I had served in Sangin and another lieutenant to cover the mortgage. For the first time in my life, money was not an issue. Andrea and I spent that summer just being young, in love, and married. We ate Mexican food. We went to the beach. We explored Southern California. I still spent time in Sangin in my mind, but I was present and vulnerable with her. It was not something that came easily to me. I am always waiting for the next shoe to drop, for the next betrayal to come. But I loved her and I loved feeling loved and I could not see feeling any other way.

I had arrived at a place for which I had been searching my entire life. I had only experienced it once before, on realizing I belonged to the infantry. In recognizing that, I found the culture in which I belonged, one where aggression and simply refusing to quit could mean more than innate talent. I found part of myself that I always had been looking for. I found a way to fill the holes left by childhood betrayals and disappointments. But the infantry alone could not make me the man I wanted to be. That required adherence to Christ's teachings and being vulnerable enough to give, and receive, love.

In marrying Andrea, I had made myself a more complete human. Combat gave me a deeper insight into myself and my reasoning, but it took things that Andrea gave me again. Combat is about destruction and ugliness, and though I could do it and knew I might be called upon to do it again, I wanted to be a good man, a loving husband, and father. I needed to become a man worthy of having her in my life.

Now I know my inability to throttle down on return from Sangin was normal, a common response for people who spend extensive time in uncommon circumstances. At the time, the military hierarchy paid lip service to mental health, but there was no culture that encouraged anyone to come forward with the feelings and emotions that result from deployed experiences. The prevailing message was just the opposite, the predictable result of a spartan mindset and a decade of nonstop warfare conspiring to make normal feelings, feel abnormal. At the time, it was just a steady feeling of being unsettled, of trying to appear to enjoy being where I was while my mind was somewhere else entirely, somewhere normal people don't even want to be. It was an almost impossible juxtaposition. I wanted everything I had at home with Andrea, but I was also a war junkie looking for a fix. Like any junkie, it was a dirty secret that only I thought I was successfully keeping.

At Lima Company 3/5, we were beginning the second six months of training required to cohere as a unit before joining the Fifteenth MEU for another six months' preparation under their organizational command. Only then would we board ships for six more months, which of course I was not excited about. That made me a prime candidate for a recruiting poster I encountered at the base gym one morning that advertised the First Marine Division's elite First Reconnaissance Battalion. Looking at pictures of Recon Marines leaping out of airplanes, kicking down doors, and locking out of submarines fifty feet below the ocean's surface, I thought, "I want that challenge, I want to fight again."

I KNOW WHO I am not. I am not the smartest, or the strongest, or the fastest guy in the room. But the infantry also taught me exactly who I am: someone relentless in the pursuit of an objective. It is a compulsion as much as an ethos. I simply do as much work, as hard as I can, for as long as I can, until a task is complete. Then I look for another task. The infantry is a job tailor-made for me, it is a force that values relentless, dedicated, and sometimes mindlessly repetitive work. It is a culture where a willingness to accept pain in the pursuit of the mission is absolutely a virtue.

Within the infantry, reconnaissance is a subset mission for which the Marine Corps specifically trains, organizes, and equips selected Marines into separate, smaller battalions. Reconnaissance Marines are trained to work in small teams and use advanced methods of mission insertion, execution, and extraction like free-fall parachuting and combatant diving to execute missions clandestinely. Additionally, Reconnaissance Marines may be tasked to perform raids upon specific targets to create gaps for larger follow-on infantry forces to exploit.

All those sexy missions pictured on the Recon Battalion recruiting poster went out the window in Sangin. There, Recon Battalions essentially became smaller infantry battalions with more operational leeway. They were conducting essentially the same operations I had been in Darkhorse. But Recon Battalion Marines were tasked to use their advanced training to go a little farther, a little harder, and a little deeper into Taliban territory than the infantry battalions. They were asked less to hold territory and more to push farther into it to reveal truths about enemy strength and disposition by engaging them in fights. They frequently operated as platoons or as companies, but in Recon either element was about half the size of a corresponding unit in an infantry battalion. Quality over quantity, aggressive operations, and a chance to again command a platoon sounded exactly like what I wanted. But first I had to pass the Reconnaissance Physical Assessment Test, and there were nearly one hundred officers from across First Marine Division with whom I had to compete for only two spots.

Especially after Sangin, I was wholly secure in my status as someone who could deal with the worst the world has to offer. But looking around at the other one hundred officers on an early morning in November 2011 left me a bit anxious. I thought of a quote from Heraclitus of Ephesus: "Out of every one hundred men, ten shouldn't even be there, eighty are just targets, nine are the real fighters, and we are lucky to have them, for they make the battle. Ah, but the one, one is a warrior, and he will bring the others back." I really hoped I was the one, or one of the two, anyway.

A voice boomed through a loudspeaker held by one of the test proctors from First Recon Battalion.

"The Reconnaissance Physical Assessment Test will commence with an enhanced Marine Corps Physical Fitness Test! In addition to a maximum effort set of dead-hang pull-ups and a maximum-effort set of crunches, both executed in a time limit of two minutes, you will execute maximum-effort sets of push-ups and of sit-ups, both within a time limit of two minutes. That will be followed by a maximum-effort three-mile run executed in shorts, T-shirt, and running shoes! Are there questions at this time?"

Those were all pretty standard events for the officers gathered in the dark. At the sound of a whistle, half of us executed specified exercises while the other half counted. Then we switched. Recon Marines circulated among us, taking names and scores. If any officer there did not hit the minimum required, I was unaware of it. Most of us went well beyond the usual Marine maximum as the clocks wound down and the whistles blew. Again came the voice through the speaker: "Gentlemen, we will now move to the three-mile run. Please line up on the road between the orange cones." This was usually my event. I had run three miles in 16:30 a month before. But that was before I broke my prohibition on running marathons.

Outside of a pickup truck and some guns, external displays of masculinity or tough-guy status are not my thing. I have never wanted to run a marathon. I don't find validation in spartan races. So when Kate Kelly, sister of my friend Rob, who had been killed in Sangin, called asking me, "Hey, do you want to run the Marine Corps Marathon?" I was quick to answer, "No."

Then she asked, "Do you want to run it in honor of Rob?" I was again quick to answer: "Yes."

I had had plantar fasciitis that came and went through training for, and running, the whole thing. Most days, to compensate, I would stop, massage my foot for five minutes, and stretch it out. Some days, I just pushed my run to the next day. But on the day we were to run twenty-two miles, my buddy Kolbe Grell and I had eaten our pasta, we had our little belts with all the gels and water, and we'd been dropped off in Newport Beach, California, with a plan to run back to San Clemente. A mile in I said, "Give me a second," followed by "I'm just going to run through it," followed by "I can't even walk anymore," at mile thirteen. We called for a ride back to San Clemente. But I was running for Rob and there was no backing out so off I went to Washington, D.C.

I ran the marathon in four hours. My left foot was so painful I considered it a blessing when it went numb at mile twelve. I ran like a pirate with a wooden peg leg for fourteen more. Ten days after the marathon I still couldn't put any weight on my left foot. The next day I got a cortisone shot directly into the arch of my foot. The day after that was the Reconnaissance Physical Assessment Test. The cortisone shot takes approximately forty-eight hours to take effect, making it both incredibly painful and totally ineffective, given that I was only twelve hours into forty-eight and about to step off on three miles that I needed to run as fast as possible among one hundred superfit men attempting to do the same. I was in pain from the first step. I was used to being at the front of the pack on a three-mile run, but now I was just hanging on, trying to pass some of the competition without too many of them passing me. I crossed the finish line at 21:30, in the last thirty or so officers to do so. That was not going to get it. Especially since we were headed to the pool.

Recon Marines swim for a living. I don't. Standing on the side of the pool at Camp Margarita, wearing camouflage utility trousers and a blouse, with a slow run already on record, I felt as if Heraclitus would not have much good to say about my chances. The loudspeaker boomed.

"On the command enter the water. You will enter the water and hold on to the side of the pool. Without pushing off the side of the pool you will commence a five-hundred-meter swim in a time limit of no more than twelve minutes and thirty seconds." Following that, I knew I had to tread water for twenty-five minutes, tow another officer for fifty meters, and then do a series of crossovers, which are alternating swim sprints on the surface and then the bottom of the pool, which left me nearly hypoxic. I passed, but I was clearly one of the least capable Marines in the pool.

I had little hope stepping into the culminating physical event, a twelve-mile run wearing a fifty-pound rucksack followed immediately by two back-to-back completions of the obstacle course. Nonetheless, the absurd level of competitiveness I am both blessed and cursed with would not allow me to stop. A half mile into the run, my foot seized. I limped the remaining twelve miles and crawled over the Marine Corps obstacle course twice. I was the absolute last Marine to complete the physical events. All that remained was an interview.

I walked into a room and was immediately confronted by two Marine gunnery sergeants and two captains. They peppered me with a series of scenarios that usually drove me to an ethical, rather than tactical, answer. They were good questions. They were also questions that I had a lot of experience with, courtesy of Sangin. My answers came not as, "Here's what I would do," but rather, "Here's what I did." I walked out confident in my performance but knowing that in the physically rarefied

Marine Reconnaissance community, confidence was not enough to get me a second look for a job with a 2 percent chance of being hired.

We all gathered outside to wait for the results. I listened to my peers pick apart their own performances, identifying all the ways in which they could have failed. I didn't need to; I could review the entire day and see my countless failures. The door to Recon Battalion headquarters building opened. The captain and gunnery sergeant who emerged gave us all their appreciation and sincere congratulations for our attempts. They expressed regrets they could only take two of us and then called off the first name.

He wasn't a surprise. I had gone to OCS with him and he was a tiny jackrabbit of a human who had led every physical event that day. I was curious which of my peers would be next. My own thoughts were already turning back to what an MEU deployment as Lima Company's executive officer would mean. Andrea would be happy I would not be attending the months of reconnaissance training: the Basic Reconnaissance Course; Marine Combatant Diver Course; Basic and Freefall parachute training; and Survival, Evasion, Resistance, and Escape school. Then the major called out, "The second selection is First Lieutenant Thomas Schueman."

My mind did a double take. I looked around for another Tom Schueman. Nope. It was me.

As the parking lot emptied, the Marines administering the RPAT brought me into their office and told me, "Look, your last three-mile PFT run was a 16:30. You couldn't run today; we obviously know you're injured. What we want are guys who have brains and your interview was definitely the best." I was still stunned, but thought of First Platoon and Zak and the experiences they had given me that got me there. I was going to be a platoon commander again and, ideally, take those Marines to combat. Perfect.

THOUGH I HAD passed the recon screening test, I had several months before I could depart 3/5 and become a member of First Recon Battalion.

First Recon was slated to deploy to Afghanistan in December 2011, but I would not join the battalion in time to make the deployment. I was disappointed to miss a chance to head to Helmand with First Recon, but I was excited to start an advanced training pipeline that included a lot of adventure. By the time the battalion returned from Afghanistan, I would be done training and qualified to lead a recon platoon in one of the companies. But as often happens, "needs of the Marine Corps" intervened.

Because of some shuffling, the First Marine Division headquarters called upon the battalion to provide an officer for a year of recon advisor duty with an Afghan recon company. I became that advisor. I would leave 3/5, join First Recon, and then be detached to commence training to deploy with the advisor team in February 2012. Finally, in July, the advisor team would actually deploy. Happy first anniversary, Andrea!

I was torn. I was deeply saddened at the thought of anything disrupting what was truly wedded bliss. I was disappointed at being delayed in attending the recon training schools I had been excited about. I was already missing what would likely be one of First Recon's last deployments to Afghanistan. Yet I was secretly thrilled at getting back to Afghanistan barely a year after leaving, even if it was in an advisor team made up of people from multiple units. With the Afghans taking the lead, I knew the Taliban would be stepping up their game and I very much wanted to play again.

Before that, Andrea and I planned to reset the official date with a formal wedding in Chicago in July. It was beautiful. I was in my dress blue uniform, she was gorgeous in her bridal white, and we were surrounded by people who loved us. When I look at pictures from that day, I see the pure love I had sought and found with her during that year. I see myself having become the man I actually wanted to be. I was living the life I had always wanted, the one I had always run from subconsciously, believing it wasn't an option for me, that it was just another way to get hurt. By opening myself to that risk over the preceding year, I had found a level of happiness with Andrea that had eluded me for my entire life. I finally found a way to stop creating chaos as a means of finding normalcy and instead created normalcy as a means of escaping chaos.

Three months later, in July 2012, I was back in Afghanistan, trying to help lend order to a place that seemed to be coming to exist only in chaos. Knowing I would be largely on my own, I thought of Zak and the brotherhood we had formed in Sangin. I had not heard from him and had no way to get in contact with him. I half-wondered if I might run into him, still in Helmand, working his magic for another platoon. I hoped he was home in Kunar enjoying a well-earned peace.

PATROL BASE LONG Beach was a triangle-shaped Afghan National Army (ANA) island outside Marjah, another Helmand Province hotspot, seventy miles southwest of Sangin. The February 2010 battle for Marjah had been the largest joint operation of the war and was heavily publicized by ISAF as the biggest operation since Fallujah in 2004. The fight went on for five days before the ANA raised the Afghan tricolor flag over the bazaar in Marjah. But two and a half years later, Marjah was still a Taliban shithole. There was a Marine battalion there, with responsibility for the town and the area around it. My job was to help advise the Afghan recon company at PB Long Beach several miles away from the next Marines.

PB Long Beach was as austere as it could possibly be. With a seventy-five-foot base and one-hundred-foot legs, the tiny outpost made FOB Inkerman look plush. With regard to amenities, Long Beach offered the purest expression of what it means to be a grunt. There was no power. I had only what I brought on my back; a cot, a pallet of MREs, and a pallet of water, all located under some cammie net and open to the sky.

The ANA lived at the peak of the triangle. My co-advisor, our small security detail, and I occupied one of the base angles. The third corner, immediately adjacent to ours, held the ANA and our latrines. We didn't share toilets, due to ergonomics. Afghans squat, Marines sit down. Sitting on an Afghan toilet would have meant getting mud from boots on our asses. Stand or sit, it was July in southern Helmand Province and sleeping next to open shitters and a burn pit was brutal.

Just getting there from Camp Dwyer, where the main advisory team was located, had proven a challenge. The recon advisors, myself, and both the incoming and outgoing Marine staff sergeants from First Recon Battalion left Dwyer for PB Long Beach in a convoy of armored vehicles. I was riding in the gun turret of one. The second, with the two staff sergeants, went off the track and hit an IED. Both staff sergeants had to be medevaced before we even reached PB Long Beach, a fact that was immediately an issue when, two hours after I first arrived at Long Beach, the ANA were suddenly in a frenzy, rolling out of the patrol base in their unarmored Ford Rangers.

I grabbed the ANA platoon commander who spoke only a bit of English and said, "What is going on?"

"Our soldiers hit IED! You come?" he replied.

I was required to have eight Americans to leave the patrol base. Without the staff sergeants blown up hours before, I had seven, counting me. Rapport is everything in advisor duty. Refusing to go out and help on day one when people were dying was not an option. I went by myself. When we got there, there was little to do but clean up the body parts. I thought, "Oh shit, here we go again." I had to imagine the Taliban laughing and saying, "We almost killed Schueman on his last day in Sangin District. Let's welcome him back to Helmand and let him know how it is in Nad Ali District!"

For all of Long Beach's austerity, what was most glaringly, immediately missing was someone for me to talk to. Training had certainly made me aware of the loneliness faced by advisors. Being confronted with the reality of it was altogether different. With my staff sergeant co-advisor medically evacuated on the first day and not due back for weeks, the only other Americans were a small, rotating security detail of six infantry Marines from the battalion located in Marjah. I was effectively alone, without a Staff Sergeant Henley or any squad leaders to talk to. Most glaringly, I was completely dependent on translators for everything I needed to accomplish with my Afghan counterparts and there was no Zak.

Of course, I had interpreters. They rotated in and out of PB Long Beach from other advisory teams. But in the summer of 2012, "green on blue" incidents, in which Afghans increasingly turned on and killed Americans, were on the rise, making rapport and clear communication even more critical. I needed Zak, who could not only translate for me, but could *speak* for me. More than that, though, I needed a friend. That made me miss Zak even more and made my two-week, mid-tour rest and recreation leave even sweeter in November 2012.

Redeployment is always strange, always an adjustment. It feels like slamming on brakes at a red light and sliding into the intersection a little too far. You wonder if everyone is looking at you. Did they see you have to back up to get behind the appropriate line? R&R leave is even stranger, because you know you will be headed back to war in two short weeks. You can't fully let down. Nonetheless, the week Andrea and I spent on Catalina Island having a long-overdue honeymoon was a dream time. The second week at home with her and my friends was one where I was able to just be thankful to be alive and married to someone who made me feel loved. Then it was time to go back.

It took a long time to get back to Patrol Base Long Beach. I was stuck at Kandahar Airfield for a week. It had only taken me forty-eight hours to get to Kandahar Airfield. In just that time I noticed a shift in the tone of Andrea's emails. Then I was stuck at Camp Leatherneck. I had nothing to do but email her and work out. The closer I got to Long Beach, the colder her emails got, as if the farther I got away from her physically, the further she felt from me emotionally. I emailed her maid of honor see if she knew what was wrong, but I was too distracted by the impending return of regular combat to focus on nuance in her reply.

While I was gone, the green-on-blue attacks had become so prevalent that I was not returning to PB Long Beach. I was reassigned to advise an Afghan battalion, or *Kandak*, commander at another base, FOB Geronimo. Living with an Afghan battalion commander meant the living got a lot better, but I was still basically by myself. More so given the fact that I increasingly felt a gulf opening between my wife and me. I never felt unsafe, as I always felt my rapport with my Afghan partners was my safety. But I still missed the implicit communication I had with Zak, the ability we had to play off one another, and the assurance that I understood my environment that came with Zak's insights.

The revolving interpreter relationships I had were very transactional, almost push-button efforts where I grabbed the interpreter and said, "Hey, I've got to go talk to the ANA, come with me. Here they are, now talk." There was no relationship that could inform how I spoke or how I understood what was said to me. As Americans were dying for imagined slights and miscommunications with their counterparts, it was a major vulnerability I felt keenly in the biggest operation I accompanied as an advisor.

During a clearing operation, the ANA were facing stiff opposition from the Taliban when one of the companies hit an IED. The Afghan Explosive Ordnance Disposal team headed out in their vehicle in response and *they* hit an IED. The *Kandak* commander sent his single wrecker to help both. When the wrecker driver went to rig both trucks for towing, the ANA *Kandak* sergeant major told him he'd done it incorrectly and stepped in to help and *he* stepped on an IED. The explosion left him missing the lower half of his face.

I'm the most squeamish guy in the world. I can't watch gory movies, I can't watch a medical show on TV, but it's not a problem when it's actually in front of me because I've had a lot of training and the real-life experience of Sangin. The ANA commander looked at me and his eyes said, "Save this guy." I called in the medevac helicopter as I told the corpsman how to treat him. We went through every emergency step possible to keep him alive and breathing, including a cricothyrotomy, in which we used a scalpel to create an opening below his Adam's apple through which to insert a breathing tube. Our best efforts could not change the fact that he was aspirating so much blood from his missing jaw that there was little chance.

The whole time, I was battling Americans on the other end of the radio telling me, "This is an Afghan problem, find an Afghan solution."

I looked around at heavily armed, distressed Afghan soldiers and said, "There is no Afghan solution for this one."

From the radio they came back, "Drive him up to Lashkar Gah."

Again, no. "Not going to work. Send the bird!"

The entire time I was fighting Americans to save an Afghan who was almost certainly doomed, more of the ANA soldiers were gathering around us. My counterpart, the *Kandak* commander, was growing increasingly desperate to save his sergeant major. "Captain Tom, please save him!"

I looked at him and the truth came out: "I'm trying! They won't send the helicopter!"

He looked so sad as he asked, "Why?" He knew why and I felt so low, as if I had betrayed a friend.

Finally, after I had put everything I had on the line to get the helicopter, the pilot called, "I'm sixty seconds out."

I was wholly unprepared to land a helicopter in a good landing zone at that moment. I didn't have a good LZ; I had a dusty courtyard full of hazards cloaked in darkness. Everywhere I looked through my night-vision goggles there was a hazard to landing aircraft.

Through my headset I heard the pilot's voice, shaking with the vibration of his rotor blades, "I'm thirty seconds out."

Then the corpsman told me, "Sir, he's dead."

It was a moment in which I needed Zak so badly to defuse a potentially volatile situation. Zak would have intuitively understood what I needed. He could have truly interpreted my words, my body language, the Afghan's body language, and the background chatter by excited and saddened ANA soldiers and said what was needed, without my direction, to mollify the grief-stricken soldiers before they took out their frustration on the only Americans they could actually see. Zak would have known the right things to tell me and the things I needed to say.

If nothing else, had Zak been present, the death of a key member of the unit could have been a moment in which my partnership with the Afghans could have been strengthened. As it was, I just hoped to survive it.

If I waved the helicopter off because the sergeant major was dead, the ANA were going lose it because they thought, "Helicopter equals survival." In the "Summer of Green on Blue" that posed a real risk to me and the Marines and corpsman with me. Conversely, if I landed a helicopter in a dirt courtyard and hurt or killed the crew and destroyed a helicopter for a dead man, I would be unable to forgive myself. It was just another lose-lose proposition in Afghanistan.

I called the pilot of the Marine CH-46: "The only place that I know that you can land is the ground right in front of me."

And that's exactly where he landed. It was terrifying, but we loaded a dead man on a helicopter and flew him off as pure performance for the ANA. Had Zak been there, maybe I could have forgone an option that put us all in danger.

WITH TROOP DRAWDOWNS continuing, by March 2013 I was ordered back to Camp Pendleton three months early. I was happy enough to head home. I sensed that something was no longer right in my marriage, and nothing much was happening in Marjah. The Afghans were taking over the fight, such as it was, but many days seemed to find me sitting with them and watching the Afghan version of *Deal or No Deal* on a TV run from a generator. I had become qualified as a Joint Terminal Attack Controller before deploying with the advisor team so, still looking for a combat fix, I sold myself to the Marine battalion as an extra JTAC as they pushed into an area called the "Hornet's Nest." The battalion sent me to Weapons Company 1/1 who were in an area called Trek Nawa, looking for the kind of trouble the ANA didn't seem interested in finding. But that wasn't so much a mission as it was combat tourism.

While I was gone, one of my lieutenant roommates, the one I served with in Sangin, had moved out. He had left the Marine Corps, and even though he and I had once been close, I wasn't sure what his next step would be. He had left our condo and Southern California, but I still called and texted him to check on him. When I did, he was inexplicably cold and distant. Andrea was treating me the same way, but I was so grateful to be home and safe and married that I subconsciously chose to ignore the fact that she didn't even seem to want to be in the same room as me. I wanted to heal and be happy. I'd had enough war. I wanted love but talking to my wife felt like I was in Afghanistan without a translator.

Things did not get better that spring. Andrea worked as a preschool teacher while I was deployed. At the end of the school year, in May 2013, she announced she missed her friends, hated the Marine Corps life, and was going back to Chicago. I was blindsided. I couldn't understand it. All I wanted was to save my marriage. The Marine Corps was suddenly a distant second. I became a hermit. I worked and then went home to call her to beg her to come back to California.

Lee Shinn, a mentor, sat me down and said, "What is most important to you? Is it being a Marine or is it having your wife and building your family?"

I was quick to respond, "It's having my wife and building a family."

He looked me in the eye and said, "Then if the reason that she isn't with you is because she wants to be back home, get out of the Marine Corps and go home."

It was devastating, but I called Andrea and said, "I'll get out of the Marine Corps. We can go home. We'll figure it out."

Her comeback was a gut punch: "I still don't want to be with you."

I moved from desperate to angry. "Okay. Well, then what the hell is the real reason?" She had no real answer.

Over the summer I convinced her to come back in August. That lasted long enough for her to go to work for a day, quit, and go back to Chicago in September. For reasons both biblical and related to my childhood, I really did not want to get divorced. Andrea and I commenced marriage counseling with her parents. I don't think she wanted to save the marriage, but she was living at her parents' house, so we tried that throughout the summer, without much progress.

When I finally asked her, "Is there anything I can do?" the answer was no.

With the summer to think about nothing else, I had to accept her decision. I wanted to be a husband and a dad. But the way I had conducted myself in the years before Sangin was not consistent with someone serious about that. The first month of combat in Sangin forced me to think, "You have not lived in a way that is congruent with the values that you espouse, with the goals that you want to accomplish. You want to be a dad. You want to be a husband. You need to grow up and be a man, because you're not doing that currently."

I agreed to a divorce. I didn't want it, but I didn't see any way out of the cycle we were in. To my shock, Andrea rejected the idea of a divorce. Maybe it was my acceptance of my culpability in our problems, but she came back to California from Chicago in October 2013.

Things felt normal again till late November 2013, when Andrea happened to leave her laptop open on our bed. Maybe I shouldn't have disturbed it. Maybe I should have left well enough alone, but I was still hurt and angry and suspicious. Just deciding not to get divorced doesn't fix all the ills in a relationship. A simple search showed me all I needed to know: the emails; the pictures; the reason things turned so strange before I was even really back in Afghanistan after R&R leave. She'd cheated on me with our former roommate, the Marine I huddled with under my poncho liner to keep him warm on a long, cold, wet night in Sangin; the guy I'd held while he'd cried over his own divorce.

For a brief moment I became every man I had spent my life not being. I screamed at her. I cursed her. Then I got back inside my carefully constructed box and told her I was going to a hotel and for her to be gone when I got home the next day. When I got to the hotel, I called my former roommate whom I'd served with and told him I knew what he had done. I didn't want an explanation, just an admission. He owed me that much.

It had been a long, painful slide into the ugliness at the end, Andrea and I were both too tired to fight for, or against, each other anymore. Andrea went back to Chicago with us both planning to follow through with the divorce this time. She didn't want it. She was beyond apologetic and repeatedly begged me for forgiveness through tears. Likewise, I felt a divorce was failing myself, Andrea, and God. But I was devastated by things both of us had done and said and, deep inside, I saw the end of our marriage, and my betrayal by a man I had fought beside, as validation of everything my childhood taught me: never fully trust anyone, never take off all your armor, always be ready for betrayal from anywhere. *Semper Fidelis,* my ass.

13

Night Letters

ZAK

IN 2014, PRESIDENT Karzai was leaving office. He had been president since the Americans installed him thirteen years before. If Afghanistan was going to have a chance to become a modern nation, we had to have fair elections to replace him. The Taliban wanted to disrupt the elections to demonstrate the weakness of the central government. In the weeks before, they were attacking government targets every day. They also attacked civilian targets throughout Afghanistan, particularly in Kabul. The Taliban intentionally chose the softest targets. In Kabul, Taliban murderers massacred civilians just for living their lives at the Serena Hotel and again at a popular Lebanese restaurant. The attacks were not justifiable in any military way. They were slaughter designed to intimidate the Afghan people and keep them from voting. It did not work.

In Kunar, people at the polling stations had to hide from Taliban gunfire. But when the fire stopped, they still stood up and voted. Inside and outside of Afghanistan, people saw photographs of Afghan women proudly holding up their fingers, stained with ink to show they voted. The successful completion of the April 5, 2014, election was a great victory for the government, but many Pashtuns would not agree to a non-Pashtun president. The results were contested. That meant there had to be a second election in June. That second election was full of the usual corruption that follows every official function in Afghanistan. Tribal politics are so powerful that even the Taliban worked to see that the Pashtun, Ashraf Ghani, beat the Tajik, Abdullah Abdullah. Ghani became president. Abdullah Abdullah became the chief executive. Even the president could not get a job without people cheating for him.

American president Obama ignored the problems of the second election and used the general success of the process to announce that America would be withdrawing from Afghanistan by 2017. Until then, Obama planned to leave 9,800 troops in the country to advise and assist Afghan forces. The remaining Americans would be spread throughout Kabul, Bagram, Jalalabad, Kandahar, and Khost. But even those troops would be gone by 2017. That news was nothing surprising to me. I had not seen an American in Asadabad since January 2014. Obama's announcement was just another sign that I should look for somewhere else to go. Deewa was pregnant and I needed to get somewhere my growing family could be safe. Unfortunately, like everything in Afghanistan, that required money. I did not have it. No one I knew had the kind of money we needed.

I graduated from the Tanweer Institute of Higher Education in November 2015, the same year my son Subhanullah was born. I was so proud of my son, and I hoped that earning a bachelor's of business administration would help me take care of him and Deewa the way I wanted to. Working for the government was a common source of employment in my village because the mountainous terrain limited how much land most families had for agriculture. For most of us, farming was more about survival than commerce. The Afghan government was trying to modernize, so I frequently checked the website that announced government jobs. When I saw an advertisement for an office and information manager in the Kunar office of the Ministry of Agriculture, I knew I would be perfect. I applied online and received a receipt to take the hiring examination.

I arrived at the Ministry of Agriculture for the examination with hopes I would be able to improve things for my entire family by earning a good job. I joined a line of applicants. Even in Afghanistan, I could not believe what I overheard as we waited to approach the desk where the chairman of testing for the Kunar Ministry of Agriculture waited to give us our tests in return for our receipts.

"You may as well go home," a man with a graying beard said to a younger man, about my age and just in front of me. "This one is already decided."

"Do not worry, old man. I have a pocketful of Afghani."

"Not as much as that man over there," the older man said, gesturing at a well-dressed man, again near to my age. "He is the one who will come on this posting. I saw the stack he paid."

"Then why don't you go home?" the younger man said as he pulled cash out of his pocket.

"I am here already. Where else would I go? I am here because I do not have a job." The old man smiled sadly.

The exam was easy for me and I was excited to get a phone call from the Ministry of Agriculture the next day. The man on the phone said, "You did very well on the exam, but you are the number two candidate. The number one candidate was only one point ahead of you. Please try again in the future." I did not need to ask who the number one candidate was.

I continued to seek government employment online and in person, but with the Americans gone and good jobs fewer and fewer, I just did not have enough money to buy a government job or any job good enough to get a home of our own and establish Deewa, Subhanullah, and me as our own family.

I did what I always did when I could not find good work: delivering groceries to people and working in my family's fields raising crops for food and sale. I also continued to help Asadullah at the radio station, performing administrative tasks. I did not get paid, but since he had three jobs, it helped him balance the load. I helped his family as we built the house addition for them by running errands and picking up their groceries. Afghan families are very connected. Where I could make his life easier, I made my own better. But I was determined to move forward with a profession that could support us and offer Deewa and my son and our future children more than what I had access to.

In Afghan culture, a man's value is tied to how he provides for his family. If I could not do that in Afghanistan, I would have to try to do it somewhere else. I did not want to leave Afghanistan, it was my home, but in early 2016 I decided I would apply for a Special Immigrant Visa to try to move to the United States. Several of my interpreter friends had applied and explained the process to me. Many interpreters wanted the visa. They saw it as a reward for their service to the Americans, especially in combat. For me, it was my last option.

I didn't want to leave my family, Kunar, or Afghanistan. I still looked at the mountains and knew they were my home. But more and more I felt there was no choice. It seemed as if the Taliban were coming back, acting as a shadow government in most provinces. The Americans and the Afghan commandos had done a good job of killing them. Now, with the Americans mostly gone and the commandos sent to fight wherever the regular ANA was overcome or quit, the Taliban was exploiting the weakness and corruption of our leaders to move back in.

Kunar felt like a less violent Sangin. The government controlled Afghanistan in the day, but the people knew the Taliban would come at night. They charged taxes and tried to tell us what we could and could not do, even in areas like my village, where the people clearly preferred the government. Even if they did not take power entirely, I heard people talking about them having a seat as part of the Afghan government. I felt like shouting, "Does no one remember the five years they controlled the country? Does no one hear the threats they make now?" Even if they did not kill me, to me an Afghan government that included them was a failure. They were everything I rejected; all I endured Sangin to fight against. Perhaps we could not defeat them, but I would not accept them.

THE UNITED STATES Congress created the Special Immigrant Visa (SIV) for Iraqis and Afghans in 2006. When they made the law, it created two programs to allow differently qualified people to get an SIV. One program was for Iraqis and Afghans who experienced "an ongoing serious threat" because they were employed directly by the United States government in some way. In 2016, I was sure I qualified for that option. Other interpreter friends continued to get night letters directing them to turn themselves in to the Taliban for punishment or else expect death. Sometimes they were form letters. The Taliban was issuing death sentences so fast they did not even personalize them. I knew of an interpreter who awoke to find a grenade taped to his front gate and connected to a tripwire. It could have killed him. It could have killed his wife and children. The Taliban did not care. On Facebook, the Taliban posted pictures of the bodies of interpreters they kidnapped, tortured, and killed.

But under the SIV direct employment option, I had to get a recommendation from a senior supervisor for "faithful and valuable service." That was a problem. I had worked for the United States government, but I was not sure how to contact anyone I worked for. I only knew nicknames for the Americans I worked for at FOB Wright and I only had one email address to try to contact them with. Also, I needed approval from the chief of mission, the U.S. ambassador. That was a very difficult obstacle for most people. I did not have any idea how to reach the U.S. ambassador.

The SIV process was even harder for translators and interpreters because when the United States Congress made the law creating two programs, it only allowed the Department of State to issue fifty SIVs each year for Afghan and/or Iraqi translators/interpreters. I do not understand why the number was so low. The program for U.S. government employees allowed for accepting thousands more applicants than that, but actually had more complicated requirements. It made me wonder if the lawmakers and the State Department really wanted any of us to come or whether it was just a law to make people feel good. There were people like me all over Afghanistan, battlefield interpreters who risked everything to directly support the American efforts. Now thousands of us were competing with one another to save our families and have the kind of lives in America that we had tried to make for ourselves and all Afghans in Afghanistan.

Fortunately, many Americans who served with people like me in Iraq and Afghanistan called their lawmakers. They raised their voices and made the politicians offer amendments to the law. The law increased the number of interpreters/translators potentially granted an SIV each year, into the thousands. Still, I was only one of nearly twenty thousand Afghan interpreter/translators petitioning for an SIV. None of us knew if Congress would agree to continue the increased SIV numbers from year to year. Despite all that, a program for Afghans like me, who worked directly with U.S. forces as translators and interpreters, seemed like the most direct and most easily proven path.

Congress created the SIV program, but the U.S. Department of State managed and approved it. The State Department website said the SIV petition process was supposed to take nine months. Within that period, I could expect that if my petition was approved, I could begin preparing to submit the paperwork required for me and each member of my family. We would need copies of passports; birth certificates; and civil documents like awards or my letter of recommendation from the Kilo Company commander. I would also need employment verification letters from the human resources sections of MEP and the U.S. government. Once that was successfully completed, at the end I would have an in-person interview at the United States embassy in Kabul and hopefully receive approval to go to the United States with Deewa and Subhanullah.

According to the U.S. Congressional Research Service, at the time I submitted my petition, 3,349 Iraqi and Afghan interpreters and their families had been resettled as lawful permanent residents of the United States. In the same years, 49,034 Iraqis and Afghans who worked for the U.S. government went to America. It did not matter which program I used; the clock was ticking louder for us every day.

Under the program for translator/interpreters, I also needed a "favorable written recommendation from a general or flag officer in the chain of command of the United States Armed Forces unit [I] supported" and a cleared background check from the same general or flag officer. I did not know any American generals. The only ones I saw in Sangin came on one-day visits. It had been almost seven years since I was there. With the Americans gone from Kunar, I did not know how to get a letter from a general. I submitted my SIV petition to the U.S. embassy in Kabul anyway in 2016.

American applications required full names. Many Afghans go by only one. Americans always want to know your birthday on any document, but many Afghan people don't know their birthday, especially if they were born during the Taliban time. They may know a general season, but sometimes people don't even know the year in which they were born. Taliban government paperwork wasn't standard from place to place and it never asked for birthdays. Of course, before the Americans came, we used only the Islamic calendar anyway. Because of all that, the most common birthday in Afghanistan is January 1. It is easy to remember in a country that never counted birthdays before 2001. My birthday is January 1, 1990. It is kind of a joke in Afghanistan because so many people celebrate their birthday that same day. In a place without birthdays, birth certificates are even more rare. These kinds of misunderstandings caused delays in a program designed for people running out of time.

There were real security concerns for the State Department to manage. Americans came to Afghanistan in 2001 because of 9/11, not because they wanted to make sure I could go to school. The Department of Homeland Security's U.S. Citizenship and Immigration Services (USCIS) agency was responsible for screening applications for security risks. It was fair to balance the challenge of trying to screen out people who were a security risk among the thousands of people petitioning to go to the United States against the time required to do so. In 2011, USCIS said it took them three to ten days to process an SIV case. They also said they did not need more people to do the job. By the time I applied in 2016, they said the problems in this part of the process were corrected, but processing time had increased to forty-seven days per SIV package.

Thousands of cases stacked up as we waited to find out if my petition would be approved. Fortunately, after I submitted my petition, the SIV office told me they would accept the human resources verification and recommendation letters from my actual U.S. employers, Mission Essential Personnel and the U.S. government, instead of requiring a letter from a general. That meant I did not have to find a Marine general who was probably in America and did not know me to ask him or her for a letter. I had contact information for MEP and the U.S. government. I was relieved, until I actually tried to contact them.

When I tried to reach my former employers from Sangin, I found that the email address for Mission Essential Personnel was not for my actual supervisor from FOB Jackson. It went to a general mailbox. I had no other ideas about how to contact him. I had to start over, establishing myself with MEP by contacting anyone I could there. The U.S. government man from FOB Wright had given me an email address when I was terminated in 2014. He told me to contact him if there was something he could do to help me. But when I tried to contact him two years later, the email came back. It said there was "no such address." I had a picture of myself with him and another interpreter, but I only knew the Americans I worked with during that time by their nicknames. I knew what U.S. state he was from, but without knowing his name, I had no idea how to find him.

The hopelessness was creeping in. I had fought in Sangin. The people I had worked for in Kunar made me even more of a target for the Taliban. I had done everything that America and Afghanistan asked of me. I did all the things America required to qualify for the SIV, but I had no way to prove it. I felt like I had lost more than two years of my life.

Of course, I did not put all my hopes on the SIV. As I petitioned for the visa, I kept looking for a job to take care of my family. Nothing worked out. Walking home from another job that someone with more money bought with a bribe, I began to think about my life. I spent seven months with First Platoon of Kilo Company, Third Battalion, Fifth Marines. We fought almost every day in the most violent district of the most violent province of my country. The Marines became my brothers there. I went home when they did. The difference was that I lived with death threats at home, too, because I hoped I could help build a better Afghanistan by working for the U.S. government for almost two more years. Hope? I felt hopeless.

I began to feel I had been lied to. I wondered if the war and the SIV petition and my work at FOB Wright were just another form of the corruption I always expected in Afghanistan. The U.S. government made promises to people like me. I began to think that maybe American government promises meant the same nothing as Afghan government promises. Maybe the only winners were people who grabbed everything they could and ran away.

But then I thought about First Platoon and how we helped each other. I thought about how when things were very bad, Lieutenant Schueman always found the answer. For him nothing was hopeless. If he was in a fight, he stayed in it. I needed to talk to him the way I did in Sangin. I found him on Facebook and sent him a message in 2016.

Lieutenant Schueman was surprised and very happy to hear from me. He told me to please call him Tom even though he was Captain Schueman now and commanding his own company, just like Kilo Company. I was very proud to learn of his promotion. I explained the situation in Afghanistan. He told me he had been back to Helmand since we were together and understood things were not going the way we had hoped they would. I explained the problems I faced. I had a recommendation letter from his boss in Kilo Company from 2011, but I needed Tom's help to get the human resources and recommendation letters. My SIV case could not move ahead without them. I did not actually realize how important it was because I did not know that the threats to kill me were no longer just threats. Someone tried to do it.

IN 2016, THE threats to others became personalized to me, too. They arrived in the form of night letters, left on our gate or nailed to a nearby tree. The night letters named people whom the Taliban threatened to kill, people who they knew, or thought they knew, worked for the

Americans. They explained our "crimes" and usually demanded we turn ourselves in to the Taliban for punishment. Mine was among them. They left off the threat to come and get me and hurt my family if I did not turn myself in, but I understood what they were saying. The threats came often enough that they were never a great worry to me. For most of us, it was just the price of seeking a better life. They did make my SIV petition feel even more urgent, though.

That year, soon after I contacted Tom Schueman and filed for my SIV, I got sick. Then sicker and sicker. I thought less about immigrating to America and more about just surviving through another day. Soon I could not get out of bed. My stomach was so painful, I was bent over. I also had pain in my back. I was throwing up constantly. I could not do anything but lie in bed. Finally, in early March 2016, my family took me to the hospital in Asadabad. No one there knew what was wrong with me. The doctor there told us to go to the hospital at Jalalabad, so my uncle and my oldest brother took me there. Jalalabad Hospital is more advanced than Asadabad. They put me in a tube for a magnetic resonance imaging test, which showed that my pancreas was severely inflamed and releasing poisons into my blood.

When the doctor came to talk to me, he asked me, "Mr. Zaki, have you worked for the Americans?" The answer was yes, but I did not give him details because I did not know if I could trust even him, a doctor in a hospital. I partially dodged the question.

"I worked to help Afghanistan become a more modern nation that cared about its citizens."

"Mr. Zaki, I need to tell you that the Taliban are poisoning people in hotels, juice shops, and restaurants in Kunar. People who have worked for the Americans and the Afghan government are getting sick and dying of similar symptoms. It looks like someone has tried to kill you with a poison. Your pancreas has been extensively damaged. You are not the first case I have seen. You are the first case I have seen still living."

It didn't seem impossible. The Taliban was clear they wanted to kill people who fought them alongside the Americans. They had threatened me specifically and called me a traitor to Islam. To know they actually took steps to kill me was the final confirmation that I had could no longer safely remain in my country. The Taliban was coming back and they would never allow people like me to live. If I did not have a family, maybe I would have stayed and fought. But I could not sentence Deewa and my children to possibly die beside me. There was not a lot of time to think about it at the time, though. The doctor looked serious as he said, "We need to pump your stomach immediately."

When they were done pumping my stomach, he gave me enough medicine for a month and said, "If you do not get better with this, come back after a month."

The medicine gave me no relief. I continued to get weaker over the month I took it. My parents and brothers took me back to Jalalabad. The doctor again gave me medicine and told me to try another month. But this time he said, "If you do not get better with this medicine this time, we will take out your pancreas."

I did not improve. Every day was torture. I did not know if my SIV would be approved. I had no work with which to support my family. I was in pain.

Only the doctor in Jalalabad kept me alive. My pancreas was infected and was being destroyed. The condition became dangerous. My blood pressure was dangerously low. Because of that, blood was not getting pumped everywhere it was supposed to go and my body's organs suffered from the poor circulation. After no improvement, a month later they took out my pancreas. I was in the hospital for a week before going home.

Even though I survived, I had a long recovery. The doctor told me it would be at least a year before I was recovered. He told me to go home and rest. I could not work much during that time because I could not walk far or carry things, so I felt bad for not contributing more to my family. The incision on my stomach was always at risk for tearing before it fully healed.

I have a huge scar on my stomach where the doctor cut me open. Without a pancreas, I have to take medicine for the rest of my life to manage my blood sugar. I am now diabetic, so I have to avoid sweets and manage my diet carefully. If I eat beef, I have a lot of stomach problems. The Taliban never got me in Sangin but they found me at home and gave me an injury I will deal with for the rest of my life. They also slowed down my ability to petition for the SIV because I was distracted by almost dying.

I had made no progress on my SIV while lying in the bed. I still did not have human resources employment verification letters from MEP or the United States government. Then Tom contacted me on Facebook in November of 2016 and said, "Zak, if you need any letters of recommendation or any help with your SIV application just let me know."

I knew I could not disappoint him if he was trying to help me, but I did not know what else to do. All I needed was the human resources verification letters, but with America leaving Afghanistan at the end of 2017, my chances were less every day. It was just not possible to get anyone's attention except Tom.

I continued to communicate with the State Department's National Visa Center for the rest of 2017. The answer was always the same. They needed the HR letters. My answer was always the same: I had no way to get them. I had my identification cards from my employment. I had pictures of my time with the Marines and the U.S. government. I had employment paperwork. But because I did not have two more pieces of paper signed by an American that said that I worked for them and when, my life and those of my family were in more and more danger. I sent a message to Tom and told him I did not know what to do. I was out of ideas. I was afraid I was running out of time.

Tom Schueman was at a dead end, too. On August 27, 2017, he posted to Instagram, "[M]y friend Zak was my interpreter during my first deployment to Afghanistan. He risked his life to support my Marines. His life is still at risk seven years later as the Taliban threaten him for his service to the United States. Zak fought to preserve life, liberty, and the pursuit of happiness on behalf of a country he had only read about. I can think of no one more entitled to share in the American Dream. I am at an impasse and not sure how to help him. Anyone that can offer assistance please let me know."

Tom's Instagram post felt like an acknowledgment that there was no way out. Whether we were going to be allowed to go to America or not, I still needed to support my family. In 2018, I found a job teaching English to Afghan children. Even with the Americans leaving and the Taliban making more gains in the rural areas, people still believed that speaking English mattered in the new Afghanistan. I hoped they were right. That same year, as cellular service and internet use continued to explode in Afghanistan, Taliban threats came by text and social media. Late in the year, they got my number, and phone calls started coming late in the night.

The first time they called me, I immediately thought back to talking to the Taliban on the radio in Sangin, telling them to come find us so we could kill them. Then I was surrounded by brothers with rifles and machine guns. We had helicopters and mortars. I wanted to respond as I had then. But I did not have a platoon of Marines now. I had only one and I was counting on him to save my family's lives.

14

Mean Peak

TOM

COMMAND IS CONSIDERED the zenith of any Marine Officer's career, and rifle company command is often the job an infantry officer most fondly remembers. It is generally the last point in an officer's career where he or she is a tactical leader able to exert direct, personal influence on Marines as they execute actions upon the battlefield. I took command of my own, Lima Company, Third Battalion, Fourth Marines, at the Marine Corps Air Ground Combat Center in 29 Palms, California, in January 2016.

In the early winter of 2014, I had received orders to the School of Infantry (SOI). At SOI, I became the director of Combat Instructor School, where I immersed myself in the details of infantry operations while surrounded by the best combat instructors at the school of infantry. I had Caleb Giles and the Lopez brothers who would eventually become some of my best friends, Merry, Garcia, Bazzone, Kent, Marroquin, and Crites. SOI also offered a predictable schedule that gave me time to process the dissolution of my marriage before accepting the 24/7 expectations of rifle company command. I did a lot of professional introspection during that time. In particular, I realized I am best suited for a profession where being aggressive and bold in the relentless application of brute force carries the day. So when I took command at Lucky Lima, 3/4, I felt like I was the lucky one.

Company command was everything I wanted, save for the fact that combat was essentially over for Marines by 2015. As a result, 3/4, including Lima Company, would be deploying not to Afghanistan but to Darwin in Australia's Northern Territory in April 2017, as part of a standing contingency element called Marine Rotational Force–Darwin. If I am honest, the work-up for it, rather than the deployment, was my favorite part of command. I was blessed with amazing staff noncommissioned officers like Leatherman, Cardenas, McCoy, Lara, and Hudson. My officers, Levine, Parker, Standeford, Clark, and Johnson, were equally superb. I was also fortunate to have peers like Justin Gray, J Adams, Nick Thompson, and our Operations Officer, Kevin Fallon. Our training took me back to my favorite place on earth, the Marine Corps' Mountain Warfare Training Center (MWTC) at Bridgeport, California, where my time commanding Lima Company confirmed what I already believed to be the truth about who I am.

In college, I loved working at the Costco tire shop because I had to wrestle unyielding objects into conformity with my will through pure physical force. That's what being a grunt is. Ignoring the stunning natural beauty of Bridgeport, viewing it purely as a problem of mountain warfare, assaulting a mountain is just changing a truck tire, writ large. It is an obstacle to be overcome. A mountain will not move for you, so you have to figure out how to move upon it and accept that no choice you make will be easy. You have to do the work.

Such was the case when Lima Company found itself in the battalion's final assault of the last exercise. My task, as Lima Company's commander and the battalion's main effort, was to move in trace behind another company until they established a defensive position, then pass through their lines to assault and conquer an objective located on the top of Mean Peak, one of the harshest pieces of terrain MWTC offers. It was exactly the kind of assignment I relish: "Here's a hard target; it will resist your attempts to overcome it. You are expected to deal with that and succeed."

Just as had happened when I was a lieutenant in Kilo, 3/5, the snow set upon us in a blizzard of historic proportions midway through the assault. Mean Peak, where I was to move nearly two hundred Marines to the top and plant the Lima Company guidon, required a twelve-mile movement just to get to the base. We pushed for a day straight. I did not sleep before we reached the lead company's defensive position, through which we were to conduct what is known as "passage of lines." After we cleared their lines, we would be on our own for the remainder of the assault.

The wind and snow were blinding as we moved into the interior of the other company's position. I was exhausted.

I turned to my executive officer, Lieutenant Sam Johnson, and said, "I need to sleep for thirty minutes. Can you keep getting everybody ready to move into whatever needs to happen?"

Sam knew that the company needed me to sleep as badly as I needed to. He simply said, "You got it, CO," and stepped off to make preparations to move the company up a mountain in a whiteout.

I lay down on the ground and threw a jacket over myself. Sam woke me thirty minutes later. I was dazed and foggy but I was soon up and ready to move. Then I heard the voices breaking through the hiss of a radio: "All units cease training, conditions are too unsafe. Go to ground."

Every unit in 3/4 went to ground and dug into the snow to get out of the wind. I thought back to the Costco tire shop. Then I flashed to The Basic School and obsessively digging a chest-deep fighting hole. I thought about times in Sangin when reason dictated a withdrawal and I yelled at First Platoon, "Press the fight!" I looked at Sam and smiled as I said, "We're assaulting Mean Peak tonight."

Sam knew me better than to even question the wisdom of the decision. The wisdom was plainly questionable. That was as obvious as the fact that I was going to persist, though both reason and regulations dictated that we stop. I had been ordered to go to ground. But in my mind, there was no stopping. I would not stop until Lima Company, 3/4 planted our flag on Mean Peak. Nobody, not weather, God, or the Mountain Warfare instructors known as "Red Hats" for the hats they wear to identify themselves, was going to stop me. And so we continued to push up Mean Peak. My nearly two hundred Marines and I gained

two thousand feet of elevation as we moved. We were not following a
path; we were scaling the shale-covered side of a mountain, at midnight,
in a howling snowstorm.

Again I flashed back to Sangin. I thought about nearly drowning
several of my Marines because I decided we needed to cross a river
when all signs told me not to. I had almost killed some Marines for
no reason and I lost important equipment. Now I was on the side of
a mountain with Jonathan Parker, one of my lieutenant platoon commanders who was particularly physically fit, trying to find a path to
move the rest of the company up into the assault. We finally had to dig
into the side of the mountain for shelter. While he and I were digging
on that side of the mountain, reason broke through. I looked at him and
said, "This is unsafe. We are never going to be able to bring the whole
company through here." Then I put that thought away and we pushed
on to find a path. The two of us had a Red Hat Mountain Warfare instructor with us who kept saying, "Sir, you need to stop. It's a weather
issue. You can't keep pushing. It's a safety violation." I ignored him. I
outranked him. He couldn't make me stop.

My radio spit forth a static-covered transmission. I had Marines
missing at the base of the mountain. One of them was Sergeant Sup-
palla, a bright and talented squad leader in whom I had plenty of confidence. I wasn't worried. The Red Hat looked at me again.

He was serious to the point of insubordination. "Hey, sir, you're done
here."

I wasn't interested in his opinion. "That squad leader will do what
makes sense, he's not going to die, he'll do something smart."

Now he was getting angry. "Sir! You cannot continue to push if you
don't have full accountability of your Marines."

I pointed at his chest in the dark. "Gunny, I have accountability of
my Marines. I know that I don't know where four of them are. I also
know they're somewhere down at the base of the mountain. We're all
accounted for. We're pushing."

We were so close to the top. We had moved a company of Marines
twelve miles to the base and then thousands of feet up Mean Peak in the

dark and howling snow. We were arrayed in the assault position, the last place we would occupy before swarming over Mean Peak like wolves bringing down an elk. I knew that the Marines tasked to "defend" the top of Mean Peak were all going to be in their sleeping bags. All I wanted to do was take the objective, figuratively bayonet the defenders in their sleep, plant the Lima flag, and then get on the radio to call in, "Lima Company has Mean Peak secure, no casualties, no problem." Victory. Game over.

The Red Hats called in their officer, a fellow captain, because I wouldn't listen to the gunny. He said, "Tom, you need to stop."

I was doing a leader's reconnaissance to finalize our plan for the coup de grâce. I looked directly at him and said, "We're not going to stop."

I had my assault force ready when the Red Hat captain handed me a radio handset. It was Kevin Fallon, the 3/4 operations officer and the number three man in the battalion. He spoke slowly and directly to me: "Tom . . . Mean Peak is secured. Lima Company has secured the peak. Objective secured. Do you understand." I finally stood down. I was dissatisfied, but it was on record. Lima Company had secured Mean Peak.

I know my competitiveness can be genuinely obnoxious. I have deep pride of ownership in everything I do. That competitiveness drives everything and it's definitely been a key to any success that I have had. I want to be the best. When I was a platoon commander there was just no doubt in my mind that First Platoon was going to be the main effort for Kilo Company in any operation. I just can't stop until that is the case.

For good or for ill, every mission I get is a no-fail mission. And that is exactly how I felt in early 2016 when I got a message from my brother Zak, telling me he needed help.

I WAS SURPRISED and thrilled to hear from Zak but immediately felt guilty that I had not been in touch with him since Sangin. We had shared so much there. It seemed impossible that so much time could pass without us speaking. Nonetheless, when you're as deeply connected to another human as Zak and I are, it is easy to slip back into the rhythms of an old relationship. Zak was the same professional, kind, and deeply respectful person when he reached out to me as he had been then. Freed from the constraints of the lieutenant and interpreter relationship, we could just be friends. But I still had difficulty convincing him to stop calling me "Tom Sir" as we chatted back and forth via Facebook, catching one another up on our lives. He was so clearly proud of his family, so obviously a father who loved them, and would do anything required to protect them. Hearing him describe his family reminded me just how much I wanted the same thing for myself.

Zak had commenced a petition for a Special Immigrant Visa to come to the United States. He had tried to make a life in Afghanistan. He had been hopeful that the changes promised in 2001 would be there for the family he was creating, but the reality on the ground made that look increasingly like a collective fantasy. The Taliban was roaring back. They seized control of the city of Kunduz for two weeks in 2015. Only U.S. airpower and advisors on the ground gave the Afghan National Security Forces the ability to regain control then. Things were not getting better and critical American support was on track to be gone by 2017. Helmand, where I had spent a year and a half of my life, was in free fall. Civilian casualties were on the increase as well. The blunt truth is that the Afghan military was less discriminant in ensuring the safety of civilians during combat. Making matters worse, the Taliban was specifically targeting civilians for indiscriminate mass attacks and assassinations, with people like Zak at the top of the list.

Though the wars in Iraq and Afghanistan had almost nothing to do with one another, despite tenuous claims made in 2003, another development had taken hold that did give them commonality. By 2015, the Islamic State of Iraq and Syria had established a branch in eastern Afghanistan, an area ISIS called Khorasan. ISIS-K was vicious in a way that made even the Taliban nervous. The two were actively at war with one another. Even though the Taliban employed attacks on civilians, especially through the Pakistan-based Haqqani Network, with their depthless appetite for blood, there was at least some debate about whether it was appropriate to do so. ISIS-K expressly advocated for mass casualty attacks on civilians. The ISIS savagery that pulled America back into Iraq in 2014 was exploding in Afghanistan by 2016. To Zak, and to me, it was another clear sign that Zak needed to get out of the country.

Zak's pursuit of a Special Immigrant Visa was also a sign to me that all the corners our leadership kept telling us were on the verge of being turned were, at best, the product of irresponsible optimism. Less charitably, they were lies designed to keep the wheels of war turning. A soldier in Iraq said, "I hate war, but I love combat." I understand that completely. There is no feeling like fighting for your life against people doing the same. I know that's not a universal feeling. Maybe it's not even a healthy one. But there is an entire Marine Corps predicated on that truth being universal enough to fuel recruiting and I won't be the one to deny it resides in me.

There is a separate truth to that, though. The war I fought was both overwhelmingly stacked in my favor and something I affirmatively chose by volunteering for the Marine Corps and the infantry. Any advantage Zak had in the war that came to his door in 2001 arrived by association with me and left with his job at FOB Wright in 2014. By 2016, as America increasingly retracted and Afghanistan devolved, Zak was not only removed from the benefits of his association with Americans; he potentially bore the full brunt of repercussions for placing his faith in his nation and ours. Zak's loyalty had always lain with his nation. If he had reached a point at which he no longer held hope in that future, I knew the war was lost. Victory now was whatever it took to honor the commitment made to him in return for his service to an idea in which I was beginning to wonder if anyone ever truly believed.

I looked at the U.S. State Department website explaining the Special Immigrant Visa process. Zak had almost all the requirements for a translator's SIV. All he really needed was a human resources letter from the company that employed him in Sangin, which was Mission Essential Personnel, and the human resources people who employed him for the U.S. government in Kunar. That seemed like it should have been easy, but neither organization was responding to him.

A peer once told me of myself, "Your skill is bending things to your will, making things succumb and acquiesce so that what you want is imposed, and it's through just determination and resolve and relentlessness." Now that Zak and I had a clear obstacle to progress, I had an objective. If an intractable bureaucracy is like a mountain of regulations, fine. I know how to deal with that. Tires are heavy, mountains are steep, and snow is cold. I thought, "We're fucking assaulting Mean Peak tonight."

I should have had more humility.

WHEN ATTACKING ANYTHING, it is best to identify its key elements and look for a weak spot. The Marine Corps predicates its entire warfighting philosophy on moving fast, bypassing positions of strength, and attacking identified weaknesses to create gaps for follow-on forces to exploit. I started by looking at the State Department process to receive an SIV to see if I could shortcut it in any way. I assumed there had to be some good-faith, reasonable means by which anyone who had risked death beside United States Marines would get some special consideration. I would make a few phone calls and this would get solved. What I learned about the SIV process is that it was less like *scaling* Mean Peak and more like *moving* Mean Peak.

For Zak to get an adjudicated SIV involved fourteen steps that sent the package back and forth between at least six different U.S. agencies spread between the United States and Afghanistan. After many reports about atrocious SIV program administration, the U.S. Congress told the State Department to start reporting on the SIV process every fiscal quarter as part of the Afghan Allies Protection Act of 2009. Four years later, in 2013, an amendment directed that SIV application review was to be completed in nine months. In April 2014, processing time for Afghan SIV petitions was 287 *business* days. I'll save you the math, but that's longer than nine months. By January 2016, around when Zak started his application, it was 293. By September 2017, average total U.S. government processing time for the Afghan program was 906 *cal-*

endar days. Three years after the State Department said it would get better at their job, they took 125 percent longer to process an SIV petition. State and Homeland Security blamed the increased processing time in 2017 on their completion of 250 cases still pending from 2015.

I simply could not believe that anyone could accept that poor an execution of their duties. It seemed simple to me. Congress said if you did something, you would get something. That seemed like a clear commitment. But for some reason, it wasn't. I did not know why and, as the ranks increasingly closed in defense of the process, the reasoning became so opaque I had no real hope of finding out. Whatever the reason, I knew we were fighting uphill.

15

They're Coming for Us Again

ZAK

AT THE END of 2017, I was still trying to get a Special Immigrant Visa and move to the United States. I did not see how it was going to happen. I had lost a lot of time recovering from acute pancreatitis after I was poisoned. I had been in the process for almost two years and on August 22, 2017, I got another email from the U.S. Department of State telling me that my SIV application was incomplete for lack of the verification letter. I sent an email to Tom about it, but I felt bad. I knew Tom was working as hard as he could to help get me and my family to a safer place, but he was an important commander. I knew he was busy. There was nothing else required besides the verification letter, but I didn't know any other ways to try to get it. And even if I got it, the processing times other people experienced made it seem there was no chance that my visa could come before Afghanistan fell apart.

In 2017, the security situation in Afghanistan was becoming more frightening. Up until his last months in office, President Obama had kept to his timeline, and by the spring of 2015, American forces had decreased to 9,800 troops, mainly advisors and the people working to support them. There were about 6,000 of our European partners remaining as well. All of the Americans and Europeans were then spread across the country at the main bases: Bagram, Herat, Mazar-e-Sharif, Kandahar, Jalalabad, and Kabul. Very few of either the Americans or

Europeans had been actually fighting the Taliban, the Haqqanis, or
Al Qaeda since 2015 but they did provide critical support to the Af-
ghan National Security Forces as they fought to take back the district
capitals the Taliban periodically seized. Some Afghan units, like the
Commandos, fought hard. The nation was very proud of them. But the
Ministry of Defense exhausted them by putting them at the front of
every hard fight in Afghanistan, especially in Helmand Province. No
one believed that the government could keep fighting the Taliban if
American air support and drones went away. The arrival of the Islamic
State in 2015 had made that even worse.

The Islamic State fighters were savages from Iraq and Syria who
brought what they learned there. They taught it to Afghans who did not
think the Taliban was extreme enough. People in the Arab countries
funded them with more money than the Taliban could pay. Without
any jobs available to young men, that became attractive to some people
who would not have joined otherwise. One of the Taliban commanders
in Kunar pledged his allegiance. Another in Sangin did, too.

The Taliban are horrible. But they did not invent suicide bombing.
It was not a part of Afghan culture to kill yourself to kill others. Al
Qaeda taught the Taliban and Haqqanis that. When the Taliban were
in charge of Afghanistan, they did inhuman things as part of administering punishment for the things they said were crimes. But the Islamic
State taught Afghans to use videos of torture and beheadings for re-
cruitment. They killed people in ways that seemed created by Holly-
wood. In one, they made ten villagers who the Islamic State said were
Taliban supporters kneel upon explosives and then blew them up. Then
they distributed videos of it by cell phone.

Most of the Afghans who joined the Islamic State were in Nangarhar
Province. They hid in the same mountains as the people who shot at me
in Nangalam. The Islamic State was not just fighting the Americans and
the Afghan government. They fought the Taliban, mostly around Jalalabad. I was in the hospital recovering from surgery on my pancreas. That
was all very near Asadabad. It made people there feel nervous about
their futures.

The Americans had originally planned for 350,000 soldiers and police to be members of the Afghan National Security Forces. Over 2015 and 2016, more than 40,000 were killed. Throughout Afghanistan, but especially in Helmand, police and army walked away from their posts. They left their weapons to the Taliban. In Sangin District of Helmand, where I had watched my Marine brothers make so many sacrifices, the Taliban established a training camp. They fought to within a few kilometers of the provincial capital, Lashkar Gah, before Afghan commandos backed by U.S. aircraft and drones pushed them back.

During 2016, even President Obama had changed his mind about withdrawal. He had decided that America would keep 5,500 troops in Afghanistan to protect Kabul until the end of the year. Then he decided that all 9,800 Americans would stay. Then he announced that 8,400 Americans would stay indefinitely, past the end of 2016. To us, it was good that more Americans were staying but bad that they had to. After fifteen years of war, it seemed the Taliban grew stronger every year, attacking the capitals of four of our country's thirty-four provinces, including Lashkar Gah in Helmand once more.

At the same time that Afghan troops were being pushed out of provincial capitals in Afghanistan in 2016, America was having a presidential election. The world watches American politics. It affects us all. We did not know what would happen in Afghanistan when Donald Trump became president. We knew he wanted Americans to leave Afghanistan, so when he announced that he would send 3,900 more troops to increase the total to 14,000, people were surprised but happy.

Then President Trump made a speech on August 21, 2017. He mentioned a possible future political settlement that might include the Taliban in the government of Afghanistan. The people of Afghanistan knew the Taliban did not want to be part of the government; they wanted to *be* the government. They were still fighting the Islamic State, too. The competition for fighters had spread to Kunar. In districts within thirty miles of Asadabad, some Taliban commanders were pledging loyalty to the Islamic State; others were fighting it. Neither of them would let someone like me live in peace.

Deep inside, I did not believe that America would ever abandon us. I thought I needed to leave with Deewa and the children because of my work for the Marines and the U.S. government, but I did not think the rest of my family would be in great danger once I was gone.

BY 2018, I was recovered enough from my pancreas surgery to begin helping my brother Asadullah at the radio station. It was important work. Afghan radio gave a voice to the people. By playing music or talking about politics or having a call-in advice show, even about how to live in a more modern Afghanistan according to the Koran, the radio station rejected the Taliban and the kind of restrictions they forced us to live under from 1996 to 2001. Maybe people in the West can't understand what it is like to have to post a lookout to listen to the radio, to hear news of the outside world, but that was our reality under the Taliban. I feared it was coming for us again.

To be able to hear and share news of your city and your country builds a community. The free exchange of ideas and information is a basic need for humans. That is what my brother was fighting for with his work. It was what the Taliban wanted to stop if they returned to control our country. It was one of many freedoms I knew they would rob from us again if they could.

The radio station was a symbol. Radio was the only means of connecting with the world beyond our borders during the time of the Taliban. Then, Afghans had to post a lookout just to listen to the radio. Someone had to be ready to warn everyone if the Taliban were coming. By 2018, most homes had radios, even villages without electricity. Broadcasting from a plainly furnished, whitewashed room on the second floor of a cement and rock building in Asadabad, a large antenna tower outside advertising what he was doing, my brother sat in front of a single microphone and passed information that people needed to make their lives better. It was not without risk. In 2018, the Taliban were aggressively targeting journalists for kidnap and murder. Fifteen had been killed that year. In Takhar Province, three hundred miles north of Kunar,

Taliban gunmen attacked a radio station and murdered two journalists as they were live on the air. It was an effort to frighten people into giving up the freedoms that make them harder to control.

I was proud to help Asadullah, but I still needed to make my own way. Our daughter, Husna, had been born in 2017 while I was recovering from my surgery. Now I had a wife and two children. I needed to start working again to support them. In addition to delivering groceries and farm work, I began managing an English education program for boys. I both taught and managed the program.

Education was everything to me—to me an education is freedom. The access to information is basic to people in the West but it was what I had been denied by the Taliban and then by our family's lack of money. When I was able to become a student after the Taliban ran from Asadabad in 2001, it was as if I had chains cut from my ankles. I was proud and happy to be able to give that to someone else, which made teaching exciting and rewarding. The teacher in my village had meant so much to me. Watching children learn made me feel the way I felt when he had come back from Pakistan and opened his home as a school in 2002. The boys laughed and chattered coming into class in a way we never did entering the Taliban madrassah. I looked at the pride on their faces when they mastered a new word or phrase, both of us knowing they were a step closer to being able to have a life beyond that envisioned by their own fathers.

The classroom was simple: a chalkboard, long tables, and benches for the boys. I taught all ages. Some in traditional clothes, some in more Western athletic clothing, the boys were all hungry to learn. Seven-year-old boys came into my class, eyes shining, smiling and laughing, as they shouted, "Hello! How are you?" followed by reply from a seventeen-year-old with the beginning of a mustache, "I am fine! How are you?" They never seemed to tire of it. Of course, we worked on many other things, but I always knew they were nearby when I heard laughter and "Hello! How are you?" from the street outside. I smiled every time. Who knew where they might go? Kabul? London? America? An education was an opportunity to have choices. Having choices is freedom.

I was also excited to be using the language skills I had developed over more than half my life. I had not been able to use my English as much since 2014. I wanted to stay proficient. English is the language of so many places that I still hoped would continue to care about Afghans and the development of our nation. English was also critical for anyone who wanted to be an active part of developing Afghanistan's role in the world. Even the youngest boys in my class knew it—I made sure of it. I planned to do the same for my daughters.

But in Afghanistan, survival comes before satisfaction. The teaching and managing just did not pay well enough for the time required. I still had to have something else to make enough money to help feed my wife, my two children, and any family members who needed my support. We were also blessed that Deewa was pregnant again with our daughter Taqwa, who arrived later in 2018. I quit and began seeking another job.

Taqwa's coming birth was not the only good news. Tom was able to find a representative from Mission Essential Personnel to write a human resources employment verification letter. It was such a simple letter, dated December 21, 2018. It just had my name and some personal information and a statement that, "this memo has been produced to verify the current employment of the below listed individual/linguist with Mission Essential as a linguist in Afghanistan," but it meant so much to me. It proved I had done the things that Tom and I knew I had. There was one problem, though: it verified only nine of the twelve months of employment I needed to qualify under the translator SIV provision. I also still needed the government of Afghanistan to survive long enough for me to get one more verification letter to show twelve months total employment and then for the U.S. Department of State to do what they said they would do in a reasonable amount of time.

While I waited, I would just keep looking for opportunity, and opportunities to make more opportunity. In Afghanistan, people are always looking for ways to survive. That's the hard part. If you can secure survival, you can try to improve your life and the lives of your family members. I thought mobile phone usage was a way to do that.

In 2016, almost 63 percent of the Afghan population was under twenty-four years old. Twenty-three million Afghans had no memory of life under the Taliban. Almost the same number of Afghan people, 22.5 million, were connected by cellular service. Many people had multiple phones. They could not imagine life in 2002, when only 30,000 Afghans had mobile phones. Walking down the streets of Asadabad, people everywhere were having conversations with people all over the world. I believed that was a sign that the Taliban could never come back to the kind of power they had before. There was no way the Taliban could reverse the kind of freedoms that people came to expect during the years after 9/11. It seemed like a business made for the new Afghanistan—the one I'd fought for.

Providing phones seemed like a good business to build a future. It seemed like something that could survive a Taliban government if I could not get a Special Immigrant Visa and my family had to remain in Afghanistan. While I did not believe I personally would survive a Taliban government—they had promised to kill me and people like me too many times—my family's welfare meant I had to continue to plan as if I had a future in Afghanistan. Still, I had no way to start a business. A business needs money to make money and I had none.

Sitting on the terrace of our home at the end of 2018, as I held a sleeping Taqwa in my arms, Asadullah and I discussed the state of Afghanistan and our families' futures. I told him my idea about mobile phones and the future of the country. He agreed with my thoughts, then he surprised me.

"Zainullah, I am going to India to study English literature. The government is paying for it. I want to get my family somewhere I don't have to worry about their safety." I told him I thought it was a good idea, even though the thought of him being far from me made me sad. How could I not support him? I needed to leave, too, even though I did not want to.

"Since I am going," he said, "I want to give you some money to start your business."

I did not want to take the money. "Asadullah, I cannot be your little brother forever. I have Deewa and the children I am responsible for now." He turned to fully look at me. His eyes shone with intensity.

"Brother, you fought for our country. During that time, you supported all of us. The Taliban has poisoned you for that. Now you will get back to working. Take the money."

I did not want to take any more of Asadullah's money. He had been working to support thirteen people for years. I had always looked to Asadullah as a leader. It had been that way since we were small children. Nothing had changed about that. I had submitted the letter from Mission Essential Personnel to the State Department but other than acknowledging they got it, I had heard nothing new about my SIV since August 2017. I accepted his help.

The Taliban had signed cease-fires before. They had even lived by them for holiday periods, like the famous four-day Eid ceasefire in 2018, but I still did not trust them. That didn't matter. I was out of options. Even if I did not expect to live long in the Taliban's Afghanistan, I had to prepare myself and my family to live or die by my decision to support the Afghan government and the American mission in my country.

I also had to make something of Asadullah's generosity to honor all the work he had done to make it possible. Throughout 2019, I stood outside my mobile phone shop in a stall in a building full of them facing Asadabad's main market street. I entered early each day through a roll-up garage door and a flexible, locking, steel gate in the front, then stayed till late in the evening. But success was hard to find. After locking the steel gate closed at night, I sat alone and looked at the day's numbers before stepping outside to roll down the garage door and lock it. As the Taliban showed they were growing in power, people were getting more nervous and trying to protect their investments. My landlord increased my rent. The cost for inventory was already high and getting higher as the Taliban continued to push into more parts of the country and the government forces proved increasingly incapable of resisting them. The Taliban could extort the truck drivers coming from Pakistan to Kunar to bring my goods. I had to pay the difference. Normal customers were holding on to more of their own money in case they had to leave suddenly or employment evaporated with local governments. Fewer and fewer people bought phones or minutes. Negotiations seemed to indicate the Taliban would have an official voice in Afghanistan and people wondered if there would even be a functioning cellular network. If people could move money out of the country, they did. If they could secure a way out, they did.

Then, in February 2020, the United States and the Taliban signed an agreement for a peace deal that established the early terms for a U.S. withdrawal from the country by May 2021. The government of the Islamic Republic of Afghanistan did not even get invited to the discussions. May 2021 would come soon and the government, and therefore we the people, would have to live by whatever agreement the Americans had worked out with the Taliban.

I was still waiting for a visa, but I had to plan for that to fail. I began to assume I would be killed in the next five years, but like everyone else, I was willing to do anything to support my family. The arrival of my fourth child in 2020, my daughter Sadaf, made me even more desperate.

I had been so hungry for knowledge and had learned so much getting my business degree—I tried to think of anything I could do to generate business. Sometimes I slept at the shop because I had stayed open late and I wanted to be at the shop and open early to catch morning customers. But the business seemed like my SIV petition. No matter what I did, it was not going to succeed.

I had worked so hard to make a life for my wife and children, for my parents, and for my brothers and sisters still in the house. Asadullah's generosity, the hours I gave to the business, and my fears for my children made it hard to accept that it was a failure. But in 2020 I had to do that. I saw my failure as a sign of things to come in Afghanistan. If people were not seeking ways of sharing information, I believed it was a sign that people felt that planning for life in a free Afghanistan was futile. People were losing hope, but no one was willing to say it out loud yet. It still seemed impossible that the world would abandon Afghanistan, even as it appeared we were waking from a dream to a living nightmare.

I had accepted the likelihood I would die young in Sangin. Since coming home, I had accepted that the Taliban would try to kill me. I had talked to my father and brothers about supporting my family if I was killed by the Taliban. Now I accepted reality and gave up my dream of building a business that could survive the Taliban. I went back to delivering groceries and working for the family in our fields. No matter what happens in the world around us, there is always family. They are everything. I was desperate to do anything that would support them. They and Tom were why I still tried to believe I would get all of us to the United States, but I was beginning to think that the chances for that were slipping away as fast as my country's future.

I had to accept reality again. There would be no SIV. But I was not going to quit. I knew Tom would not quit. Still, mentally I prepared to die.

16

Protect No Matter What

TOM

THE YEARS SINCE my divorce in 2013 had been a time of self-discovery. I spent the first year as a hermit, staring into the dark of my condo and ruminating over my failures. I barely left except to go to work. But my assignment to School of Infantry offered me a fixed schedule and a chance to practice what Marines call gym therapy. As I got physically stronger, I transformed my exterior. But by focusing on being who and what I expected of myself, and who and what I believed God expected of me, I grew into a better person.

But for all that, by the time I took command of Lima Company, 3/4 in January 2016, I needed Andrea. Despite both of our failings, only Andrea could make me feel the way that I wanted to feel. Only she could make me truly feel loved and in love. She was the woman I wanted to spend the rest of my life with.

So, I did what I had to do. I called her.

I talked to her. I went to therapy. I decided to get fine with not feeling fine. We each unloaded all the truth about all the reasons we felt we had both failed the first time. Then I went and saw her in Chicago before I deployed to Australia. I had wasted enough time. I told her I wanted to get back together. She said yes and I smiled my way through seven months in Australia's Northern Territory.

While I was deployed, I discovered that the United States Naval

Academy, in Annapolis, Maryland, was looking for an English professor. I knew I was going to take an enforced break from the operational ranks of the Marine Corps on the back end of commanding a company. It was a normal career step to move into a job within the "supporting establishment"—the part of the Corps that serves as the oil keeping the wheels of the Marine Corps turning. Common positions for guys like me were in recruiting, as higher staff drones churning PowerPoint presentations, or working somewhere like equipment acquisition and modernization. Three years as a student and teacher sounded better than any of that. More importantly, it sounded like a stable way to reinitiate my marriage.

Was I an ideal candidate to teach college English? Maybe not. I majored in political science and communication at Loyola. There, I failed English because I only attended the first and last day of class. Then, with one day's notice from the Marine Corps, I took the GRE, which predictably resulted in a near perfect score on the reading portion and my bombing the math portion. Given those shining bona fides, why wouldn't I apply to be an English professor? If I was accepted, all I had to do before teaching literature at the U.S. Naval Academy was complete Georgetown University's two-year English graduate program in one year. Easy! I had so many regrets coming out of my undergraduate studies, summed up by the fact that I had largely ignored a chance to grow intellectually, that I was not going to repeat that mistake. It was just another Mean Peak I had to scale and I did it, applying and getting accepted for the USNA program and Georgetown alike. That it had all begun because of a Google search for "great works of fiction" while I was deployed in Australia made it almost funny. I had ended up reading Dickens's *Great Expectations* and rediscovering a love of literature just as the Marine Corps made it known that the Naval Academy needed an English professor.

My life was looking up but Zak's was clearly not. He had let me know he had health issues he was dealing with and then I had not heard from him. I reached out to check on him and his SIV application in November 2016 and found out that he still needed the same human resources verification letters I had been trying to secure from Mission Essential Personnel with only limited success. One call led to another as I climbed the corporate ladder. Each email yielded another person I should try. I was having even less success with the U.S. government people for whom Zak had worked. I had no more luck with the email address his boss there had given him than he did and his specific employer remained confusing. I had friends who had moved on from the Marine Corps into work with other elements of the government. I asked them for advice, but no one could really tell me where to look.

I GOT BACK from Australia and gave up command of Lima Company, Third Battalion, Fourth Marine Regiment in January 2018. Not long after, Andrea and I got married, again, at the court house in Joshua Tree, CA. Six months later, I left 29 Palms and reversed the cross-country journey I had made eight years before. I wasn't that same young man anymore. He wasn't all gone, but I had left much of him in Afghanistan, for good and ill.

Now I had a lot to think about. I had been to war twice. I had married, divorced, and was about to remarry the same woman. I had commanded a company of Marines and the Marine Corps had seen fit to select me for promotion to major. I was now a career Marine, on my way to spend three years molding future Navy and Marine officers.

As I crossed our nation, I visited the graves of nine Darkhorse Marines. Where and when I could, I visited their families. They were still living, breathing men to me. They certainly were to their families. As I began to turn my thoughts to training and educating future officers, I felt it critical to visit and reflect in the presence of some of the men who made me who I am: Corporal Tevan Nguyen, Sergeant Jason Peto, Lance Corporal Alec Catherwood, First Lieutenant Robert Kelly, First Lieutenant William Donnelly IV, Corporal Derek Wyatt, Lance Corporal Arden Buenagua, First Sergeant Christopher Carlisle, and Sergeant Matthew Abbate.

All but Sergeant Peto and First Sergeant Carlisle died in Sangin.

Sergeant Peto made it back to Walter Reed hospital in Maryland before he died. First Sergeant Carlisle survived his wounds the day that Zak and I watched from a distance, then got promoted, retired, and was killed in a motorcycle accident. I owed him much of my sanity while in Sangin. I needed to tell him that.

Matt Abbate was the heart and soul of 3/5. A recipient of the Navy Cross, the second-highest award for valor presented by our nation, his death in combat stunned every Marine in the battalion. He was a Scout/Sniper who spent much of his time in support of Kilo Company. He was "the one" of whom Heraclitus spoke, the one who "will bring the others back." No one was cooler than Abbate. No one cared more than Abbate. He was the author of "The Gunfighting Commandments" posted on a wall at PB Fires and now inscribed on the back of his grave marker:

THOU SHALL NEVER LEAVE WIRE WITHOUT BANDANNA CONTAINING AT LEAST 4" OF SLACK . . . IN ANY SITUATION THOU SHALL BLAZE . . . NOTHING MATTERS MORE THAN THY BRETHREN TO THY LEFT AND RIGHT . . . THOU SHALL PROTECT NO MATTER WHAT . . . WHEN GOIN' OUT IN A HAIL OF GUNFIRE . . . THOU SHALL POP THEM NUGS UNTIL THE BODY RUNS DRY OF BLOOD . . . AND LOOK HELLA SICK.

I hoped I could live up to the essential truth of his charge: keep the faith with your brothers and never quit moving forward if you have a breath in your body. I owed that same standard of duty and loyalty to Zak.

Even if I was discouraged by the fact that we had met with no success thus far, I was no more ready to give up than I had been at any other time in my life. Then I got a message from Zak that said, "[H]ave a happy life and say hi to all my friends. Now I am going to English class to teach some boys."

It seems innocuous on the page, but I understood the real meaning in his message. Zak had lost hope that we would prevail. It was more than a sign-off. It was a farewell message from a condemned man. I felt like crying, but quitting was not an option.

As I crossed the country, taking the side trips required to stand and acknowledge the hard service of good men, I mentally redoubled my commitment to a brother still living and still in danger. To have done any less would have obviously been a disservice to Zak, but also to the men who stood, and died, with him in Sangin. I felt like I had to salvage something out of the entire experience as a means of making their sacrifices, and the losses suffered by their families, mean something. As I watched the odometer click off miles, I wondered whether anyone truly cares about war but the warriors and those they directly affect.

By 2018, Afghanistan was beginning to appear to me to be a colossal footnote that Washington just wanted to close out. I felt like the American people were unaware that Afghanistan still existed. Meanwhile, I had been home from Sangin almost seven years and there was not a day that passed without me remembering some facet of that seven months. Sometimes those memories were happy. Sometimes they were fraught with sadness, though my ability to feel it was still blunted, as it had been since that bloody November in 2010. I did not know if I would ever completely get home from Sangin. Maybe I left too much of myself there with my memories of better men. But I hoped that somehow bringing Zak to America would help me come home, too.

EVEN AS I arrived at Georgetown University, I was still somewhat stunned that I had even been chosen for the program. The reality that I had the actual opportunity to spend a year studying and earning a master of arts in English literature at Georgetown University before coming back to teach at the United States Naval Academy in Annapolis was nothing short of amazing to me.

While at Georgetown, I started writing my own words as much as studying the words of masterful writers. After almost ten years, I was ready to start making sense of some of what I had seen and felt in Afghanistan, particularly in Sangin. I began looking at the process and narrative of veterans' homecoming. As part of that, I began questioning the very structure of the machine that creates the need for veterans to come home. Thinking of the pain still evident on the faces of loved ones of good men dead almost a decade, I became convinced that sharing the burden of war across the whole of society would reduce the appetite for war among people who would never fight it and save grief among what is increasingly America's warrior class. Visiting D.C.'s many think tanks, I grew adamant that we must dissolve the incestuous relationship between the Pentagon and the defense establishment, the same military-industrial complex against which President Dwight D. Eisenhower, himself a five-star general, warned. Perhaps most dramatically, I began to believe that a Peace Amendment to the U.S. Constitution prescribing defensive war as a fundamental aspect of our national ethos was wise and in keeping with both our existing defense architecture and our constitutional ethos.

My study of the classics helped, and the words of people like Shakespeare and Dickens had particular impact. Where many critics see Shakespeare's Coriolanus as a martinet, imbued only with martial talents and limited education, I felt the internal conflict of a man with such an assured sense of self that he preferred death to sacrificing his principles. I felt like I knew him. The Greeks; Homer and Epictetus spoke directly to me. But the Georgetown curriculum also forced me down some uncomfortable literary roads where real growth occurred as I explored subjects completely new to me under professors like Kathryn Temple, Sherry Linkon, Matthew Pavesich, and Norma Tilden.

Asian American Literature with Professor Christine So had the greatest impact on me. I credit her with teaching me how to read closely. That was a gift. I think the ability to read with deep attention to what is said, and what it's saying, is a bit of a superpower that she gave me. Using those skills, I also watched a panoply of war movies to better understand the messaging and presentation of veterans in motion pictures. The overwhelming narrative presented of veterans was as broken humans. I certainly was *different* postwar, but not all of those differences were negative and in no way was I broken or nonfunctioning. I started thinking about emotions and their importance. I contemplated our collective societal attempts to insulate ourselves from our own emotions: substances, pharmaceuticals, and sex, to name but a few.

As I crystallized my thoughts about the veteran's return from war, I started trying out my ideas at open-mic events in Washington. I stood in front of strangers, acknowledged my own darker emotions, and advocated sitting with sadness and anger and guilt as natural by-products of trauma rather than trying to suppress them.

In what would become a lecture at the Naval Academy, I said, "We've somehow stigmatized being sad or angry. We've somehow made it that when a veteran comes back from combat and they say that I am lonely, or I am angry, or I am sad, we say, 'Oh my God you must have something wrong with you!' And what I want to argue is that no, there's nothing wrong with that person, they're just sad! It's okay to be sad! It's okay to be angry! It's okay to be lonely! And it's not just okay to feel these kinds of things one day, it's okay to feel them in seasons! So, we've got to stop telling people as soon as they have an emotion that there must be something wrong with them!"

I encouraged service members to speak about how they felt, particularly to acknowledge when the emotions had become more than they could carry alone. I was deeply sincere and increasingly committed to my developing ideas. It felt like a matter of life and death because it was. All of it was uncomfortable. All of it was essential. After Georgetown, I was energized by learning; excited by teaching and writing. I was in a solid mental space to head to Annapolis and help create another generation of military leadership.

From the first day in the fall of 2019, I loved teaching at the United States Naval Academy. I worked with smart, dedicated faculty to teach smart, dedicated Midshipmen. One thing the USNA does incredibly well is recruit the very best our nation has the offer. The men and women who were there as Midshipmen were so far beyond where I had been at that age. They were mature, fit, and serious. Standing in front of them each day was a privilege.

As a Marine, nothing beats service in a combat unit. However, for the two years I spent teaching at Annapolis, things in my professional life could not have been much more satisfying and I was deeply professionally fulfilled. I served as the Officer in Charge for Leatherneck, a program for training and evaluating Naval Academy Midshipmen for service as Marine officers, I was a representative to the USNA Infantry Skills Team, and I mentored the Midshipmen of 20th Company. The Marine Corps competitively selected me to attend the Naval War College in residence after the USNA, another year in school offered to the top 10 percent of majors in the Marine Corps.

I had sought another, very different set of trials and had either overcome them or was engaged in meaningful work to do so. Andrea, too, was a big part of that. She and I had endured challenges, come back together, and were now where we had first envisioned being years before. Most importantly, God blessed us with two beautiful children, Amelia and Jack. These were different trials, to be sure, but the trials of fathering a young family were ones that I had craved for as long as I could remember.

Teaching in the English Department, I was given the room in front of an expert audience to test my thoughts on the hero's journey and the veteran's return. I began to truly think about how to harness the power of social media for the mental, physical, and professional health of my military community. Through an Instagram page I titled @kill.z0n3, I began broadcasting my lectures and engaging people, mainly younger service members who used social media as a primary means of communication and interaction. The kill zone is the receiving end of an ambush. I chose the name because we all get ambushed in life at some point, we all have traumatic moments. The idea I wanted to explore was how can we become more resilient in an effort to avoid the kill zone altogether, but when the inevitable occurs, how do we recover? That became even more important to me in April of 2020 when Corporal Justin McLoud lost his battle with pain medication and died of an overdose.

TWO MORE MARINES who had served under my command died by their own hand the same month Justin McLoud passed. They were not the first Marines, or even the first people, I knew to take their own lives. Suicide has been a specter haunting my life since my childhood friend's father, the man who cared enough to gift me with time at father-son dinners, hung himself in his garage when I was nine. The man my mom dated while I was in high school, who lived in my house for years, lay down on a train track while I was deployed to Australia. But the Marines were the ones who made me realize that whatever we were doing wasn't working, that *Semper Fidelis*, Always Faithful, only means what Marines make it mean, and that the *Semper* portion, the Always, must be absolute. It means being faithful *every* time. I saw that as the mandate for getting Zak out of Afghanistan. Likewise, I saw it as a mandate to be there for hurting Marines, every time. Young people die in combat, but far more were dying in peace. The number frequently cited is twenty-two veteran suicides daily. The Department of Veterans Affairs puts that at seventeen. One is too many. I cannot abide the idle admiration of problems, particularly lethal ones, and it seemed that people were content to describe and admire the problem without attacking it.

At the Naval Academy, I asked for, and received, professional help. I began to unpack and examine the emotions I brought home from Afghanistan locked in boxes. As I began to open the boxes and examine the contents, it was hard for me to contemplate how badly the people we had failed collectively must have been hurting, how sick they must have been, or what signs we had all missed. I hated to think of people I loved being alone and in pain, especially in an organization predicated on victory through teamwork. I joined the Marines out of a generalized desire to be of service. I had to find some way to provide help to people who needed it.

Part of my research at Georgetown led me to Jonathan Shay and his work on moral injury, *Achilles in Vietnam: Combat Trauma and the Undoing of Character*. Shay compares the experiences of Vietnam veterans with Homer's descriptions of war and homecoming in the *Iliad* and the *Odyssey*. I was deeply interested in the issue and it had become a fixture in my lectures at the Naval Academy. The death of Justin McLoud made it personal.

As I looked at the problem over the summer of 2020, I realized that the vast preponderance of military metal health resources are directed at physically wounded veterans or the special operations community. The bar to entry for support services was just too high for most service members. Moreover, they are employed reactively, after someone has tried to kill themselves. I also learned that most service members who committed suicide were not combat veterans and that they killed themselves in the first year after exiting service. Many people who had offered themselves in service, and desperately needed that continued community, felt excluded because their service was not spent knee-deep in spent brass and pulled hand grenade pins. I was no more ready to leave them behind than I was Zak.

The suicide problem among service members was emblematic of the broader issue of the disposability of the warrior class. Since America ended mandatory service and conscription, we have effectively outsourced the most solemn duty of the nation to a tiny percentage of volunteers. One percent of America carries 100 percent of the responsibility for the nation's wars. It is considerably easier to initiate a war when neither yourself nor your offspring will bear that burden. Like children with a new bowl of goldfish, the initial fervor for war soon gives way to other interests and the goldfish find themselves left alone in a bowl on a shelf in increasingly dirty water, ignored by the children who so excitedly demanded them in the first place.

As I thought about the issue, it was impossible not to think of people like Zak, who answered our call and found himself ignored or victimized by bureaucratic incompetence, and wonder if I had more in common with men like him than many of our fellow citizens. We were all disposable heroes.

In November 2020, I founded a service organization called Patrol Base Abbate as a means of offering community to all service members in the hope of helping them reconnect with their warrior spirit, foster community through shared interests, and rediscover purpose in service to each other, their families, and their communities. Patrol Base Abbate features a "Patrol Base" (PB) in Montana. Members work to improve it while in attendance at various retreats based upon common interests.

PB Abbate has rapidly grown to fifty chapters nationally and internationally. I felt like we were making progress getting ahead of the specter of suicide haunting so many service members. But I was still no closer to getting Zak out of Afghanistan.

BY THE SPRING of 2021, for all the joys and honors, contrasting my family's lives and the places in which we had arrived within them made Zak and his family's situation in a rapidly deteriorating Afghanistan that much more urgent for me.

President Donald Trump lost the election in the United States in November 2020. Before Joe Biden was inaugurated, Trump announced he would cut U.S. troop numbers in Afghanistan in half, to 2,500 Americans, before he left office. Biden took over in January 2021. Four months later, he announced that America would end military involvement and depart Afghanistan by September of that year. I had four months to get Zak out.

Zak was now in imminent danger. We had moved beyond "if" to "when." We had made some progress in 2018 by securing a human resources letter from Mission Essential Personnel. The letter verified Zak's employment in Sangin from September 24, 2010, to July 5, 2011. But that letter presented just another problem: the SIV required twelve months of service as an interpreter/translator. The MEP letter gave us about nine and a half months. I was desperate to get Zak out of Afghanistan with his wife and children and it wasn't going to happen on a translator visa.

During five years of his applying for an SIV under the interpreter/translator category, I had hoped that we would find the people for whom Zak worked at FOB Wright, see what they could offer in support, and make a judgment about our best chances for success under either provision. Once I found out that the State Department would accept human resources letters in lieu of a general-officer letter for translators, I felt that would be the simplest method and hoped that the almost three total years he had done translation for Americans would carry the day once we secured the FOB Wright letter. Now we were going to have to rely on the second option, that of people who had worked directly as U.S. government employees, regardless. We submitted the MEP letter anyway, figuring more support was better.

We needed a recommendation from a senior supervisor for "faithful and valuable service," along with approval from the chief of mission, the U.S. ambassador. By the late winter/early spring of 2021, it was clear the chief of mission was going to be busy as thousands of people sought to escape what was coming in Afghanistan. As Zak said to me in a message via Facebook on February 15, 2021, "you may better know the situation is worse in Afghanistan than ever. I would be lucky if I get a chance to travel to the US to save myself and my family."

It had become clear to me that the problems Zak was having securing a visa were beyond just me. I had neither the power, the experience, nor the legal acumen to move his case forward. I naively assumed that someone with immigration experience could suggest some magic waiver that I was missing. But I also realized that crowdsourcing was increasingly a means of getting things done; that social media was not just an outward facing entity. I could ask for help.

In over a year, my Instagram page, @kill.z0n3, had gained steam as I used it to discuss the Marine Corps, the military profession, and our 501(c)3, Patrol Base Abbate. I had built the page to about fifty thousand followers. I realized I could employ it to help Zak just as I had used mortars and aircraft to support us in a firefight.

On April 26, 2021, I posted an Instagram message titled, "I need your help." In it I said, "As the war comes to a close, my interpreter Zak is still stuck in Afghanistan. The time for good ideas or people who might be able to help is over. Zak has done more for this country than 99% of Americans. . . . If you are willing to take point administratively and legally to get Zak to America, please send me a note. Thanks for your support. His life and his children's lives are on the line."

The response was significant and a validation of the virality of social media. People with connections to Congress and the Senate answered, offering to use them. People from actual congressional and Senate staffs reached out. People with money offered it. People with neither asked me how they could help.

Beth Bailey, a freelance writer who had written an article about PB Abbate for the *Washington Examiner,* authored the first story about Zak's situation there on May 10, 2021. She wrote another in a military and veteran-focused magazine called *Coffee or Die* on May 19. Things started moving rapidly and in ways for which I wasn't entirely prepared and I certainly wasn't authorized.

I didn't tell anyone at the Naval Academy about what I was doing, figuring forgiveness was easier to seek than permission. Besides, what was I going to ask permission for? To save a friend's life? I'd be lying if I said I wasn't nervous about possible blowback for my public appearances, but *Semper Fidelis* means *Always* Faithful. There is no clause about "unless it's potentially personally inconvenient." As I had told Beth Bailey, "Zak and I served together; Zak and I fought together. I make a lifelong commitment to the people I serve with."

The urgency I felt to save Zak had built to the point that I needed every moment of the day to respond to the possible avenues of support opening. I was stunned at the number of people who *wanted* to help. At the same time, I was disheartened by the number of people who actually *could.* I had been trying to help Zak escape Afghanistan for almost six years. Where was the consolidated official effort? Where were the directions about what people like Zak were supposed to do? Where were the people he worked and risked his life for? What was the plan?

I had been emailing with Greg Holt, a staff member for Representative Tom McClintock of California, since December 2020. He was probably the most engaged person I had found to assist in Washington, D.C., and I really appreciated the thoroughness he displayed in keeping me updated. But there appeared to be just no way to break the logjam holding up Zak's case. No one gave a damn that I could speak to Zak's valor on repeated occasions. He didn't have a full twelve months as an interpreter for the military and, where he was employed for enough time, there was only silence on the other end when we sought help. A single form letter was going to be the difference between life or death for six human beings.

All I've ever had to rely on was a willingness to keep rolling a stone uphill. The night before I graduated from the Infantry Officer Course, we had a traditional event called Warrior's Night. There was a lot of alcohol involved when I realized there was only one lieutenant in the class whom I had not forced to tap out during hand-to-hand combat training. He hadn't forced me to, either. I knew nothing about mixed martial arts. I just refused to let anyone beat me, relying again on pure will, pain tolerance, and brute force, to bend an opposing force to my will.

My competitive spirit, and a lot of whiskey, compelled me. I looked at him and said, "Hey, man, let's go outside. Let's do this." We faced off then collided like two lions fighting over prey. We struggled for positional advantage as we rolled in the dirt, until he managed to put me in an armbar. I was not going to tap, but he also didn't apply the slow, steady pressure required to safely conduct combatives training. I heard my elbow pop when my ulnar collateral ligament tore. He heard it, too, and let go of my arm. I got up and went for him again. I couldn't even move my arm, but people had to break us up. That's why I am a grunt. I will take the pain required to accomplish any goal I set for myself or that gets set for me.

So, I put my head down and went to work for Zak. Over May 2021, as the Midshipmen were wrapping up exams and preparing to commission into the Navy and Marine Corps, I found myself talking to members of Congress during my fifteen-minute class breaks. As a Marine officer, talking to members of Congress is usually a scripted dance where everyone in uniform knows what they're going to say and why. We are conditioned to believe that little good comes out of surprise conversations with politicians, particularly ones that don't involve generals as interlocutors and managers. But I had Zak's life to save, and I had not seen much from the upper echelons of American leadership, uniformed or civilian, that convinced me they were the ones to make it happen.

Tom and his sister, Jessie.

Zainullah Zaki, 2012.

Grace (Tom's mom), Jessie, and Tom.

Tom and Andrea at the 245th Marine Corps' Birthday Ball.

First Platoon, Kilo Company, Third Battalion, Fifth Marines at Camp Leatherneck, Afghanistan.

Zak in Kabul, July 2021.

Infantry Officer Course 4-09 Palm Field Exercise (FEX), 29 Palms, California (Lieutenants Babos, Schueman, Smith, Szostak, and West).

Tom and First Squad Leader, Sergeant Trey Humphrey at Patrol Base Abbate in Montana.

Zak in Sangin, 2011.

Lieutenants Will Donnelly, Cam West, and Tom at Enhanced Mojave Viper (EMV) before the deployment to Sangin.

Kilo First Platoon at Justin McLoud's memorial service. His son, Desmond, is in the center.

First Platoon on patrol in Sangin. Corporal Justin McLoud, Tom, Doc Rashad Collins, and Lance Corporal Eric Rose with Sangin residents.

Tom and Corporal Aguilar discuss security in Sangin with residents.

Tom and Staff Sergeant Tim Henley in Sangin.

Zak teaching First Platoon Marines to skin a goat.

Tom with law enforcement professional Vern Horst after an operation in Trek Nawa in 2013.

Tom after promoting Tim Henley to First Sergeant at Camp Lejeune, North Carolina.

Zak providing security in Sangin.

Tom with one of his interpreters and the *Kandak* commander at FOB Geronimo in 2012.

Tom holds an RPG-7 rocket-propelled grenade launcher Zak found while searching a compound in Sangin.

Lima Company, Third Battalion, Fourth Marines (3/4) at Camp Pendleton, 2016.

Tom with Corporal Tevan Nguyen's son, Tevan, in Hutto, Texas.

Lima Company (3/4) on Mean Peak at the Mountain Warfare Training Center (MWTC) in Bridgeport, California.

Tom and Andrea on their wedding day.
(Courtesy of Gina Pecho)

Tom and Andrea at the Marine Corps
Marathon in 2011.

Tom with the Infantry Skills Team during a visit to the United States Military Academy
at West Point.

Tom with Jack, Marty, and Amelia.

Tom, Rob Kelly, and Cam West at Camp Wilson during EMV.

Tom with former secretary of state George Shultz at Stanford University's Hoover Institution during graduate school.

Zak and his family outside Hamid Karzai International Airport (HKIA).

Marines guarding
HKIA review
documentation in
August 2021.

Afghan citizens trying to make it through the
screening point for the Abbey Gate; most would
never make it past the wall. The tower is the site
where the suicide bomber detonated his vest.

Hamid Karzai International Airport in August 2021.

Zak near the North Gate
on his second attempt.
He was so close, but
proximity wasn't enough.

During Zak's successful
effort at HKIA. It was his
last shot in every way—
thankfully the rescue team
spotted him.

The scene outside HKIA.

Zak; his wife, Deewa; his daughters, Sadaf, Taqwa, Husna; and his son, Subhanullah, inside HKIA.

Subhanullah, Fort Pickett, Virginia.

Zak, Deewa, Sadaf, Taqwa, and Husna being shuttled to a plane at HKIA.

Wheels up: Zak, Deewa, Sadaf, Taqwa, Husna, and Subhanullah on the plane about to depart HKIA.

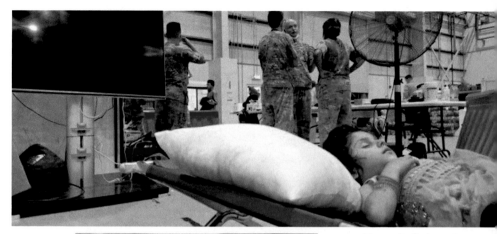

Husna in Qatar, where they were flown after their escape from Kabul.

Subhanullah in Qatar.

Zak with Sadaf, Taqwa, Husna, and Subhanullah in Qatar.

The rescue team: Major Jared Lefaivre and Staff Sergeant Alex Frogley with Zak, Deewa, Sadaf, Taqwa, Husna, and Subhanullah. (Staff Sergeant Greg Veccharino also participated in the rescue but is not pictured here.)

Worth Parker with Subhanullah and Husna, Fort Pickett.

Worth and Zak at Fort Pickett.

Zak and family on their way to their third attempt at HKIA.

Reunion in the U.S.: Zak and Tom in Minnesota, February 2022.

I sent Zak a message: "Bro make sure you check your Whatsapp. I have some very important people sending you messages to help."

And they were. I could not believe that after twenty years of war, the evacuation of our allies was coming down to direct messages from politicians and their staffers. The offices of Illinois senators Richard Durbin and Tammy Duckworth were engaged on Zak's behalf. In Congress, I had good connections in the offices of Representatives Ted Budd of North Carolina and Seth Moulton of Massachusetts. On the ground in Afghanistan, I had an avenue to Major General Curtis A. Buzzard, through his daughter, a Naval Academy Midshipman. Lieutenant General Lori Reynolds, the highest-ranking woman in the Marine Corps emailed me directly to ask if I could help with another interpreter. For all of it, what was so jaw-dropping to me was that even with the interest of some of the most powerful people in the U.S. government, without a letter, a simple administrative requirement, I simply could not move Zak's case forward.

Advocates for Afghan interpreters put the number assassinated by the Taliban over the course of the war somewhere between 300 and 1,000. As focused as I was on trying to get Zak out of Afghanistan before the Taliban could do the same to him, as they had threatened to do multiple times, I had duties as a father, husband, and Marine.

I've always wanted the obligations of leadership under crisis, at least since the Infantry Officer Course. But I could also feel the pressure building inside me. It was only June, and I wasn't sleeping or eating. I was surviving on willpower, tobacco, and energy drinks. When I wasn't leading the Midshipmen on conditioning hikes and obstacle courses or talking to them about combat leadership, I was calling congressional staff members to beg for help or answering calls from journalists. I was living the closest thing I'd felt to the constancy of combat since 2013.

On June 8, 2021, Senator Durbin talked about Zak while questioning Secretary of State Anthony Blinken. Durbin said, "I'm going to send you from the articles—Thomas Sherman, Schueman rather, he talks about the fact that his interpreter who risked his life to save this Marine Corps officer, is now at risk himself."

"I saw that, I read that article," Blinken replied.

"Can you give me your assurance," Durbin asked him directly, "as you've been asked by Senator [Patrick] Leahy and others, that this will be a high priority to protect those men and women and their families, whose lives are at risk because they were willing to stand up on our side?"

It seemed like a victory. It really was just another crack in the wall of my faith. I could not believe that a sitting United States senator could question the United States secretary of state about Zak's case and nothing would come of it. But that was the case. It was Shakespeare's "sound and fury, signifying nothing."

I felt like I was in Sangin again, surrounded by people who wanted to do great things, the Taliban outside the gates, ready to ambush us every time we came out. There were traps waiting for us, some of which had the potential to be catastrophic in their effect on the mission to save Zak. The situation was changing moment by moment, one choice mandating another. Worse, we could neither see nor predict many of them and all of them had the potential to kill my friend and his wife and four children.

JUST AS PEOPLE had volunteered their legislative connections, I started hearing from people with friends in the press. Sasha Ingber of Newsy was the first person since Beth Bailey to connect with me on Zak's story. I had gotten over my reluctance to speak to legislators. Now I was getting over my innate reluctance to speak to the press. That was a hurdle. Involvement with the media typically involves an approval process by a public affairs officer who talks to you beforehand and explains what you should and should not say. Often, they accompany you during an interview. That's actually as much an insurance policy as it is a constraint. The professional and personal dangers of saying the wrong thing and hearing about it from someone higher in your chain of command are real and potentially long lasting. It's generally easiest to just avoid the press. But after five years of institutional failure to honor a promise made to Zak by the United States Congress, one written in blood on the battlefield, I would do anything, speak to anyone, risk whatever, to help Zak.

What I did not expect was Zak's willingness to talk to Sasha and anyone else who could help. He was under an active threat of death. Putting his name and face on the international news was absolutely a desperation play that had the potential to make it easier to hunt and kill him. But that's where we were.

By late June 2021, a northern swath of Afghan provinces had fallen. Those were areas that had been reliably anti-Taliban, now in Taliban hands. Professional analysts and press commentary alike gave the Afghan government six months once American forces pulled out. Some people clung to hope that the weapons and training and money that America and our international partners had poured into the country could somehow stave off the inevitable. Zak had no such illusions.

I asked him, "What will happen if I can't get you out?"

His response was clear: "Then I will be killed with my family."

That seemed even more likely when the last American at Bagram Airfield left in the night without telling the Afghan base commander on July 2, 2021.

Handed a deal negotiated by the previous administration, who had not seen fit to involve the government of the Islamic Republic of Afghanistan, President Biden had said on June 24, 2021, "Those who helped us are not going to be left behind." He said they would be moved to third countries while everything got figured out. But if the largest military airfield in Afghanistan was no longer an option, how were the promises made to thousands of our friends and allies going to be honored? The easy answer is, they weren't. Twenty years of war on peace by a succession of presidents were instructive. There was never going to be an effort to make good on a deal struck when no one thought it would matter, and now people who counted on the United States to do the things it promised to do, the things that make the nation what people believe to be, were going to die. I stepped up my campaign.

On July 8, 2021, Jennifer Steinhauer of the *New York Times* met me at a coffee shop in Edgewater, Maryland, near Annapolis. We called Zak. Sitting outside, watching people sipping their lattes as we tried to save Zak via a teleconferencing application and the coffee shop's shaky Wi-Fi, I could not help but feel a flash of anger at the immensity of the privilege that allows Americans to live a life wholly ignorant of the kind of dangers that Zak accepted as fact. At the same time, I was so deeply thankful that my own children were safe.

We reached Zak in hiding at his brother's office. "Tom, brother, I have received another threat at my house. The Taliban told me to surrender myself or face death."

I felt the pit of my stomach go cold. As Jennifer interviewed him, I assured him, "I will keep working this for you every day and every night until we get this taken care of. I'll never forget you, brother."

It was hard to hear Zak over the sound of the traffic passing by us, but he thanked me when I told him that the GoFundMe campaign I started had raised $15,000 for him and his family. Still, I felt so impotent. How do you look at a friend and brother, so close on a six-inch screen but so far away in real life, knowing he may be killed in the next hour and be completely unable to affect it? I was so tired, I choked up. I didn't know what else to say.

That week, I'd been shaving early one morning. I'd looked into the mirror. My eyes were tracked through with red and swollen from lack of sleep. Dragging a razor across my face felt like an effort. I knew I would spend any moment in the day not actively engaged, in a fog, trying to parse out the next step on an unmarked path full of political and professional IEDs. But I thought back to Abbate's "Gunfighting Commandments": ". . . NOTHING MATTERS MORE THAN THY BRETHREN TO THY LEFT AND RIGHT . . . THOU SHALL PROTECT NO MATTER WHAT . . . UNTIL THE BODY RUNS DRY OF BLOOD." Zak is my brother and I had to protect him no matter what. I knew that if I failed to do the right thing, the thing I was ethically and morally obligated to do, I would have failed men like Sergeant Matt Abbate as surely as I failed Zak.

Staring at myself in the mirror, I thought about the lifelong bond

I shared with Zak. On the one hand, it was no different from what I shared with all the Marines with whom I fought. On the other, there was something more than the ugliness of combat that created the connection Zak and I had. In the decade since Sangin, Zak and I had become husbands and fathers. I saw the beauty in that connection that surpasses any relationship born in the stink of fear and homemade explosive. Zak and I are linked by our kids; by the common instinct of fatherhood: to die before allowing harm to come to our families. When I looked at photos of Zak's kids, a beautiful boy and three precious girls so close in age to my own children; when I contemplated what it meant for them if I couldn't get them out, I knew I would have failed my own family as surely as I had failed Zak and his. There had been enough failures in Afghanistan. I swore to myself that this would not be one more.

It was Mean Peak again. We're going up. I would find a way.

A week later, on July 14, 2021, Zak and I were guests on the *New York Times* podcast *The Daily*. Our efforts were gathering steam. More and more people were calling, asking how to help. The most important was Major Jared Lefaivre, United States Air Force.

Initially Major Lefaivre reached me on Facebook messenger and said, "Hey, I'm going to Afghanistan. I should be flying people, let me know what you need." I knew Jared from The Basic School. We were Marine lieutenants in the same company. I was chasing assignment as an infantry officer and he was already guaranteed a trip to pilot training. Through a circuitous route, Jared was now a rescue pilot in the Air Force, flying the HH 60 Pave Hawk helicopter. He had been home from Syria for forty-eight hours when he got orders to head to Kabul, Afghanistan, for duty as the director of operations for a Personnel Recovery Task Force tasked to rescue downed pilots as things got worse. He got to Afghanistan on July 15, 2021, and was at Hamid Karzai International Airport (HKIA). We talked back and forth via messenger, but now I wanted to talk to him via phone, despite a nine-and-a-half-hour time difference. The connection was weak, but I could understand him.

"Jared! Dude, it's amazing to talk to you!"

He was all business. "Tom, Bro, I would love to chat, but I have a hundred and thirty people, three helicopters, and two C-130s I am trying to run in the middle of this shitshow. What can I do for you?"

I got to it.

"Hey, are there any plans to evacuate people from Kunar?"

What he said next made me nauseous: "No, if they're not in Kabul, they're not getting out."

I paused to think. "Okay, do you know if there will be a letter or an email saying, 'Go to Kabul'?"

"I don't think so. You need to tell Zak to come to Kabul and get as close to the airport as possible."

"Okay, brother, I appreciate it. I'll get him moving."

On July 16, 2021, I called Zak and told him, "You need to get to Kabul. Do you have anyone there you can stay with?"

Zak was understandably puzzled. "If you don't mind," he replied, "I need to know for what we're going to get to Kabul?"

To stay alive seemed too direct an answer, but it was accurate.

17

Waiting in Kabul

ZAK

ON JULY 16, 2021, when Tom told me I needed to get to Kabul and explained why, I had to stop and sit for a moment. I was relieved that more people than just Tom and I cared about the survival of my family, but I knew chaos was coming. I just needed to be still before it began.

Even when you've been planning for it, the thought of leaving your home forever is overwhelming. How do you say good-bye to your family members when you know you may never see them again? How do you kiss your mother for the last time? How do you keep from forgetting the smell of her hair after you do? I sat on the terrace of our home in Asadabad, where so many of the simple things at the center of my life had happened: childhood, marriage, my children taking their first steps. They were moments as tied to a place as I was. Now I was being forced to leave them behind with my home.

For all my sadness, there was no hiding from the situation in Kunar and the rest of Afghanistan. Life for regular Afghans was bad and getting steadily worse. I was living in hiding. The Taliban's messages—which had long made it clear they were looking for me and would kill me for working with the Americans when they found me—were growing bolder and coming more often. I could not live in Kunar. I could not survive there. More importantly, my family could not.

When I had learned that the Americans left Bagram in the middle of the night on July 2, I thought about my own times flying in and out of there. I remembered how the air base felt like it could not contain all the planes and helicopters. It seemed like America's power was without end. I had placed my faith in that. I did not see then how the Americans and our government would ever let the Taliban come back. But here the Taliban were, leaving me messages at my family's doorstep during the night.

I told Tom I would go to Kabul myself and find a place to live before I went back to Asadabad for my family. I did not want my wife and children on the road between Asadabad and Kabul more than once. The last time I had come home from Kabul was 2011. Then the Taliban checkpoints were only in the night. Now there were more of them and they were out in the daylight, looking for people who worked with the Americans. It was unsafe for my children to be with me.

I had no car and no money. I just had friends. One volunteered to take me to Kabul. Another promised to loan me money to get there. I hoped to get to Kabul, find a place near the airport for Deewa and my children to hide with me, and then call to my family in Kunar to let them know. My brother Nasirullah volunteered to bring Deewa and the kids to me once I had a secure place to live so I would not have to risk the road again.

I felt embarrassed to have to ask for help. Everyone needed all of whatever they had: money, gas, food, and time. Prices for food were going up every day, gas for vehicles was getting harder to find, and the Taliban were getting closer. In our minds, they were already there. Everyone was getting more desperate. Then the friend who was going to loan me money called to tell me he could not. I had no other way to pay for the gas in my other friend's car. I did not know what I would do.

When I spoke to Tom and the reporter, Jennifer Steinhauer from the *New York Times,* on July 8, Tom told me he raised $15,000 to help us get out of Afghanistan. But the money was hard to access quickly, and I needed it now. On July 23, I sent him a message telling him that things were not going well. I had not found a way to Kabul yet. The Taliban were coming in force. They had taken control of a district in Kunar. I felt a trap closing around me.

Tom promised to wire me $5,000. When I went to the wire office to receive it, they had rejected the transfer in America. They kept asking Tom for more verification and when they finally allowed him to send the money from America, it got rejected in Afghanistan because my name was reversed from Zainullah Zaki to Zaki Zainullah on some paperwork. Many people in Afghanistan have only one name. In wealthier families, someone may have a surname, but it is not the same as the parents', as in America. It may be a tribal name, a father's name, or just something that describes you. There was no time to waste and cultural confusion was stopping me from getting to Kabul. When the wire company allowed the money to come through, Tom's bank shut it down. Finally, we got it wired.

On July 24, 2021, I found a way to Kabul. During the entire ride there I was looking for Taliban. I stayed low in the passenger seat and tried not to make eye contact with anyone when we passed through towns. Every time I saw a man with a long beard, I wondered if we would get stopped and I would be executed. I knew of at least one interpreter who had been dragged from his car and beheaded on the side of the road in Kunar. The streets were emptier than I had ever seen them—unless they were fleeing their homes, people stayed in them. At traffic intersections where they once stood on pedestals, and at checkpoints throughout the city, in the places where I had always seen Afghan Police, there were none. The visible symbols of the government of the Islamic Republic of Afghanistan were simply evaporating.

With the money Tom sent, I found a one-bedroom apartment on the fifth floor of a building behind a girls school in Kabul and paid for three months' accommodation. The building had a security guard, which was important to me. In the apartment by myself, looking out the window at the school, I wondered how much longer girls would be allowed to go there.

I called Nasirullah. "Brother, bring my family if you can." He was blood. I knew he would keep them safe.

"Zainullah, your family is my family. God willing, we will be there tomorrow."

While I anxiously waited for them to arrive on July 25, I spoke to Tom. He told me he would connect me to an American man who could help us. Tom met John Shattuck through someone who connected them from social media. Mr. Shattuck had helped get other interpreters out of trouble. He saw Tom on Instagram and now he wanted to help me.

As soon as Tom connected us, Mr. Shattuck had questions. He was sending me messages as fast as I could answer them. After feeling like Tom and I were alone for years trying to get an SIV, it felt good to have someone else helping me. He wanted to understand where I was and how many ways I had to communicate. I had a smartphone and an old cellular phone. I used the smartphone to send him a map to the apartment I rented. He told me to charge them both and make sure they always had power. He also said to stay inside and away from windows. I understood. I already was living quietly. I did not want to expose myself. The Taliban were not in Kabul yet, but they were coming. I did not know who in the neighborhood might inform on us once they got there.

The next day, Nasirullah brought Deewa and the kids to me. It was a relief to have them with me again. The children were excited by the city outside and wanted to tell me of the things they saw on the drive. They wanted to play as well. I could not ask four children under six years old not to make noise, but we did our best to keep them quiet with games and songs sung quietly.

By August 4, 2021, all six of us had been inside the one-bedroom apartment for nine days. Tom was working to make my case in the American press. He wanted me to talk to a woman named Rachel Maddow on the American channel MSNBC the next day. I said okay but first I had to get our passports to get out of the country.

When I left, just going down the stairwell of the apartment felt dangerous, like leaving the gate at FOB Inkerman. The morning was still cool, but the sun was so bright it hurt my eyes after ten days inside the apartment. Most people in Kabul speak Dari instead of Pashto. I speak both, so I said "*Sobh Bachir*," good morning, to the guard standing at the gate with his AK-47. He smiled and nodded his head in return, placing his right hand over his heart. As I stepped past him and through the gate onto the street, I felt like I was again on patrol in Sangin. I did not know what might happen or what direction an attack might come from. I only knew that this time, while I wasn't alone, my brothers were on the other side of the world.

Like in America, Afghan passports come from the post office. When I got there, even early in the morning, there were already people waiting for the door to open. They also hoped to find a way out. They all looked wary. Perhaps they were hunted, too. As we moved into the building, there were so many people we had to stand close to one another. You could smell anxiety and fear in the crowd. Some people tried to push ahead of their place in line, but others sent them to the back. Several times I thought there would be a fight. I waited quietly. I did not want to attract anyone's attention.

When I got to the clerk, he looked as if he knew what I wanted before I said it. I spoke quietly, as if I had something of which to be ashamed: "Sir, I need to apply for passports for my family."

He snorted, "A passport? It takes months!"

I did not know what to say. He saw the desperation in my eyes. His own narrowed.

"Of course, there is a fee you can pay to expedite a passport in a few days . . ." I knew what he wanted. This was Afghanistan. There is always an option.

"How much is the fee?" I asked.

"For three hundred dollars per passport you can get them in one or two days," he said.

Eighteen hundred dollars was two months of combat for me in Sangin. I knew we did not have months. We barely had days. But I knew I was not going to get passports that day no matter what and I did not think it was right to ask Tom to pay a bribe. It is normal in Afghanistan, but not in America. I went outside and called Tom. It was late at night in Rhode Island.

"Tom, they are saying it will be eighteen hundred dollars to get my family's passports. This is a bribe. What are your instructions?"

"Zak, we will pay the money! No question, brother! Getting your family out is the mission. I'll send you the eighteen hundred. Just be ready for Rachel Maddow."

I went back to my apartment to wait for Tom to tell me the money was coming. I had a message from John Shattuck telling me there were reports that the Taliban was tracking down interpreters through Facebook. He gave me some steps to take to make my account more secure so we could still communicate. After a few hours, Tom let me know he was ready to transfer the money.

I went down on the streets again. There were many more people moving in the afternoon. I wrapped my *destmal* around my face even though it was 90 degrees and walked a few blocks to the Western Union office. This time the money came through much faster. With $1,800 in my pocket, I was even more nervous on the street, but I made it to the post office. I was there for hours. It felt like the time I was lying on an IED in the middle of a firefight in Sangin, counting the minutes until I was either dead or free.

I paid the $1,800 bribe to the clerk. We both acted as if that was a normal part of applying for a passport. He told me to come back the next day.

When I got home, my youngest daughter, Sadaf, was restless. I took her in my arms and held her close to me. I whispered in her ear and bounced her on my knee. I promised her this would be over soon. There was little else I could do. We could not go out. I couldn't sleep that night. In Afghanistan, a bribe may possibly not be honored if a better deal comes along, and tomorrow rarely means tomorrow—it just means the future. It's the only way we keep hope.

I woke early on August 5 and watched the street. The lights were still shining, bathing the parked cars in a sickly yellow. There had been no streetlights twenty years before, before the world came back to Afghanistan; I wondered how many lights would be there in a year. As the sun came up, I reflected upon years of predawn prayers in Kunar. I knelt to pray, sadly hoping I had only a few Afghan sunrises left ahead of me.

I was going on American television that morning. When I tapped the link that they gave me, I was speaking to a man named Ali Velshi. He asked me, "What message do you have right now? For the president of the United States, and for Americans, about the situation that you and others like you face?"

I never meant to speak for all Afghan interpreters, but you have to do what the situation demands. "I, first of all, say to the President Biden, we are respecting you, because you are the leader of America. We were your allies on the field, and the mission of the United States, we help the mission of the United States, now is your time, our American friends, our American partners' time, to help us, and get us out of this crisis because the enemy are looking for the people who worked with Americans."

It seemed so strange. I was sitting in Kabul as Afghanistan tore itself apart, hoping people in America could help me, even as America had been unable to help my country survive itself. But no matter who saw me on TV or what Tom was doing for me, only I could get the passports.

When I got back to the post office, I again waited in line with equally desperate people. The fear stink was worse. The same clerk as before was there. He dumped out an envelope and our passports spilled out. I began going through them to verify them. My passport was waiting! In my mind, I thanked Allah and Tom. The kids' passports were also ready. But Deewa's was not there. I counted again. One. Two. Three. Four. Five. I looked up at the clerk.

"Kindly, sir, I am missing one passport."

The clerk I had bribed had no explanation. He looked inside the envelope as if I might have missed it. Maybe he thought it might appear. He looked at me.

"It didn't come to the post office."

I was angry. "Yes, I am not an idiot. I see it is not there."

"Try the main office of passports at the Ministry of the Interior."

I left with the five I had and took a taxi to the Ministry of the Interior. He didn't offer me the three-hundred-dollar bribe money back for the passport that wasn't there.

I watched the streets as we drove, and where there should have been armed police, there was no one. At the Ministry of the Interior, they checked their records. They said Deewa's passport was printed and sent to the post office. I did not know if I was angriest at the usual corruption, the usual incompetence, or the usual apathy. I knew it would not matter which one got my family killed. I went back to the apartment to tell Deewa I still had no way for her to leave. She was in tears.

The next morning, I returned to the post office. I walked a different route. I did not know who was watching and Mr. Shattuck said not to establish a pattern. At the post office, they said Deewa's passport still had not come, that it was lost. I felt as if I were lost with it.

Exhausted and out of ideas, I went outside and sat in the street. There was no way I was leaving without my wife. I sent a message to Tom and Mr. Shattuck to thank them for trying to help and to say I could not get Deewa's passport. They answered almost immediately, like they were waiting on me. They said not to worry about it, that no matter what, they were not going to stop until I was out of the country with my family. They said the passport would not matter; they were finding other friends to help us. Tom told me to be ready to talk to the *New York Times* podcast *The Daily* again the next day. They had more questions from the first time we talked in July.

On July 8, 2021, a reporter had asked President Biden, "Is a Taliban takeover of Afghanistan now inevitable?"

The president said, "No, it is not . . . [b]ecause you—the Afghan troops have 300,000 well-equipped—as well-equipped as any army in the world—and an air force against something like 75,000 Taliban. It is not inevitable."

I wondered if he read the same news I did. When I woke on August 6 to news that the Taliban had taken Nimruz Province, in the southwest near Helmand, I wondered if President Biden regretted saying it. Jowzjan fell the next day, followed by Sar-i-Pul, Takhar, and Kunduz. The north of the country was traditionally anti-Taliban but it was falling faster than anywhere else. In the streets of Kabul, there was terror over what Taliban takeover would mean. They were wild animals from the most unsettled parts of Afghanistan. How could they run a country that even educated people supported by the most powerful countries on earth had failed to save?

I was in that kind of mind when I talked to Lynsea Garrison from *The Daily*. We had connection problems, of course. As I told her, "This is Afghanistan, you know? Yeah, sometimes, you need something, it doesn't work. If you don't need it, it works."

I thought about the post office and bribes. I told her about life under a death threat. She said, "I mean, I guess I just want to ask, when you heard that, what went through your head? What were you thinking?"

I wanted to say, *I was thinking I just want to live somewhere that I am not going to die for trying to make a better life! I was thinking the fact I even have to explain it to people while my children live it makes me angry at all of us. I want to be free to read and vote and learn what I want. I want my children to go to school and my son and my daughters to have choices in their own lives and not have to leave their country to do it!* That was the truth. But what I said was also the truth: "That we're helpless, you know?"

All along, John Shattuck was working to plan my exit. He constantly sent me messages updating me about plans for my movement to the airport. I hoped John and his friends could move quickly. Kandahar, the home of the Taliban, fell on August 12. So did Herat in the west, on the Iranian border. In the east, Ghazni went for the Taliban. The Taliban were now within an afternoon's drive of the apartment we were renting. The U.S. Embassy closed down and moved to Hamid Karzai International Airport. We had known the military was leaving. But to know the American flag was no longer flying in the Green Zone? That brought despair.

Trapped in the two rooms of our apartment, I watched Afghanistan's TOLOnews on my phone and listened to the remaining radio stations narrate our nation's slow-motion suicide. My phone buzzed constantly with texts from friends and relatives seeking more information about our plans or confirmation of wild rumors. I knew the agreement that President Ghani made on August 12, to resign and leave peacefully in return for the Taliban stopping at the edge of the city until the end of August, was the end of everything we had hoped for, fought for, died for. It was the end of the Afghanistan we thought we could be.

I think the city was in shock for a day. But by August 14, the major city in the north, Mazar-e-Sharif, was in Taliban hands. Rumors were flying, despite news of Ghani's capitulation, maybe because of it.

I sent Tom a message: "Sir, so sorry. I'm okay but the situation is not good. Internet is not working. I didn't get my wife's passport yet and the bank system is also not working. I didn't get the money, and the Taliban said they'll attack on Kabul in 74 hours."

Everywhere, people were panicking. People were looking for ways out of the country. At the post office, where I waited on three separate days, people were still trying to get passports. At the banks, people were draining cash out of the vaults. Food prices were already high, but they increased overnight. Outside of Kabul, the Taliban was picking up American weapons and equipment dropped by police and soldiers. Inside the city, the police and soldiers were taking off their uniforms and dropping them where they stood. Smart ones held on to their weapons.

The next day, August 15, 2021, President Ghani left Afghanistan behind him. Nangarhar and Wardak Provinces collapsed. The noose tightened around Kabul and all of our necks. That night, after my children were asleep, Tom told me it was time to go to the airport.

18

Semper Fi

TOM

BY AUGUST 14, 2021, I had failed Zak countless times. That I had not secured a visa for him felt like a personal failure I could not justify. Zak and I had been working on it for almost six years. I was not alone in my failures. There were countless institutional failures in Afghanistan, stretching back decades. The lack of a cohesive strategy during a twenty-year war; Afghan tribalism and overwhelming government corruption; revolving-door military deployments; muddled United States immigration policies seemingly designed more for political exigency than as a reward for loyal service; and poor record keeping by pop-up contract companies grabbing plentiful defense dollars before the well dried up, did nothing to help minimize the backlog of almost twenty thousand SIV applicants in existence by August 2021.

But for me, it came down to what I said about the Special Immigrant Visa program during an interview on ABC's *Nightline:* "I think it's a very simple transaction. You serve with U.S. forces and we will provide you a visa. He served with U.S. forces; we did not provide the visa. I think that's a betrayal." The only person without blame was Zak.

However widely the field of blame for that betrayal lay, leaders must accept responsibility. I was Zak's leader and my failures were those most tangible to Zak. I was not a faceless bureaucrat in a windowless office, churning paper and computer files. My face was the one that looked up at Zak through a swirl of dust and smoke. My body was the one he stood over, holding my rifle, protecting me as I lay wounded. My eyes were the ones that saw him run two hundred yards across IED-laden ground to single-handedly capture a Taliban fighter before he could ambush us. The SIV program was a collective debt to thousands of Afghans and Iraqis, incurred by the United States. Mine was a personal debt owed to Zak. I was honor-bound through blood shared on the battlefield and duty-bound as a citizen of a government who made a promise. I was certainly morally obligated. But I felt almost legally obligated by the notion that if America makes you a promise, and I am America to Zak, then I'm going to uphold that obligation of duty, morality, and love.

Adding to the pressure of this moment was the fact that I was in the process of moving our family and starting my studies at the Naval War College, in Newport, Rhode Island. After I had finished my assignment in Annapolis in July 2021, I had to move my family, including two children under three, and one due in forty-five days, to Newport. Thankfully, Andrea, even as pregnant as she was, had the lead. She would get everything done required to move a family of five for the third time in three years, then take the kids to Chicago to be with her parents, while the movers got things to Newport. I'd planned to go to Newport ahead of them, arriving on July 14, 2021. Class was not scheduled to commence for two weeks, and I was living in a hotel on the base because our rental house was not ready. Alone in Newport, I could dedicate twenty-four hours a day to saving Zak and his family.

For the rest of July, my job—when I wasn't actively working to get Zak out—had been fund-raising, which meant bolstering our GoFundMe campaign. I had begun publicizing Zak's story with anyone who would listen. Whether it was someone's podcast recorded in their bedroom or the *New York Times,* local news or *The Rachel Maddow Show,* I said yes. I had gotten over my military-conditioned reluctance to talk to the press. As a result, we suddenly had thousands of dollars in our GoFundMe with which to operate. But actually getting money to Zak when he needed it had proven to be a challenge.

Ignoring the amazing fact that I could quickly send large sums of money from Newport, Rhode Island, to Kabul, Afghanistan, wiring money was a constant administrative nightmare. I called another interpreter I knew, a man already living in America, to have him talk me through the paperwork required to send money from the United States to Afghanistan. Even when I did get through the layers of confusion created in the translation from English to Dari and Pashto, my own bank cancelled the transactions to ensure I was not being defrauded. As a result of the repeated administrative roadblocks, I had been using the Schueman family checking account to move Zak from Kunar to Kabul at the same time I was trying to fund a move to Newport via Chicago. It would have been fair of Andrea to ask me if I was sure this was a good idea, but she knew me too well for that.

A hotel room had proven to be a lonely place to contemplate the possibilities that could come with failing Zak further. By July 25, I was running on emotional and physical fumes. I had been awake almost constantly since July 23, trying to track the progress of Zak and his family's move from Kunar to Kabul. The nine-and-a half-hour time difference between the Eastern and Afghanistan Time Zones meant my bedtime in Newport was start time in Afghanistan. I didn't have a radio operator or a platoon sergeant to take a watch, so I slept sitting up with my phone in my hand when I slept at all. The Marine Corps has conditioned me to manage operations in a certain way. Directing information flow is essential to that. But in July 2021, the only flow was Zak to me and me to the world via social media and the press. I had no ability to task anyone. There were no orders to give, no squad leaders to trust; there was just me in America and Zak in Afghanistan, both of us trying to make things happen in some semi-coordinated way.

In the dark times, especially between midnight and dawn, my mind ran to worst-case scenarios. The thought of Zak's children under threat from the same people who had tried to kill Zak and me countless times was the source of waking nightmares for me. Fixing so intently on Zak and his family also reinforced for me how much I missed Andrea and our kids. It made me feel all the more alone when outside my hotel I could see families enjoying a paradisical Newport summer.

A combat deployment, for all its dangers, offers certain luxuries. Chief among them is focus. When I had been fighting in Afghanistan, that was my only job. With survival the standard, I had been able to either ignore or hand off the mundanities of life. Now I felt like I was still fighting the Taliban, while also living the life of the suburban American dad and husband I have always aspired to be. It was physically exhausting before I considered the emotional toll my laser focus on Zak and his family might take on me and mine.

On July 24, Zak had finally arrived in Kabul and found an apartment within a long walk of Hamid Karzai International Airport. He'd rented it for three months, but I knew we did not have that much time. If Zak was still in Afghanistan when the lease ran out it was because he was buried there. Deewa and the kids were scheduled to arrive in Kabul on July 26, Kabul time. In the meantime, I connected Zak with a man named John Shattuck, recommended to me by a social media contact.

I didn't know John, but he was vouched for and had a long list of people he had assisted out of difficult situations. John had been part of a larger effort to successfully bring out an Iraqi interpreter code-named "Johnny Walker." "Walker" was the interpreter for the late Navy SEAL Chris Kyle, of *American Sniper* fame. John was ready to do the same for Zak. I needed the help. I was handling every angle of evacuating Zak myself and something was going to fall through the cracks. If John knew people who could help Zak arrange actual movement on the ground, that would allow me to handle the broader publicity, finances, and logistics of Zak's situation.

John's previous experiences had left him very concerned about security, particularly of our communications. Zak and I had been using Facebook Messenger, so John had Zak change his name on Facebook and moved us over to an encrypted communications app. With John laying the groundwork for movement to the airport, I could focus on the Hail Mary for the Special Immigrant Visa.

Zak's nine months of work in Sangin, followed by almost two years working for the U.S. government in Asadabad at FOB Wright, would more than qualify him. But the names of the people who employed him at FOB Wright remained a mystery, whether by fog of war or by design I did not know.

As the virality of our social media campaign continued, I also started to hear from people in the Marine Corps and Navy network who wanted to help us. Most of them had their own reasons, based in their own service with the Afghan people, and their views on America's obligations were similar to mine. Often they could not help whomever they left behind in Afghanistan because they had lost touch or the person was dead. Zak was becoming a proxy for a lot of American consciences.

I was sitting in my hotel room in Newport when the phone rang during a brief moment of silence. It was Navy Captain Erin Meehan, one of the members of the Naval Academy English Department who knew a retired military officer through some of her previous duties. She suggested I speak to him about Zak, assuming he would have contacts among units either already in Kabul or on the way.

I called the man she suggested immediately and said, "Sir, this is Major Thomas Schueman. I understand we have a friend in common."

He was to the point: "Yes, we do. If she says you're good with her, you're good with me. What do you need?"

That was a refreshing start after years of roadblocks. "Sir, I need a few things. I need some way to find the people who employed Zak from January 2012 to December 2013 at FOB Wright in Asadabad. He has a picture of himself with one of them. I need a letter from those people verifying that period of employment."

"Well, who *are* those people, Tom?"

I felt stupid.

"Sir, I don't know. That's the problem. Zak says it wasn't the military, but beyond that I only know it was part of the U.S. government. Zak was managing laborers and cleaning crews, not doing stuff outside the wire."

"Okay, I've got a strong network among people who might have been in Asadabad back then. Send me that picture. I'll reach out to them and start pulling strings and I'll get back to you."

I'd hung up hopeful that at least someone besides Zak and me would be sending emails into the ether.

The remainder of July had passed without much forward movement for Zak and his family. As Zak's family joined him in Kabul and the world watched the Taliban close in on Kabul, Andrea and the kids had arrived in Newport from Chicago. Andrea was eight months pregnant with our son, Marty, and had been solo parenting two kids under three for the better part of two weeks. Fortunately, the movers arrived a few days before Andrea and the kids, so we could start settling into our rental house the weekend before I began classes at the Naval War College on July 26. Thankfully, my professors at the Naval War College were accomodating, so I had time to help Andrea with some of the myriad details that come after a move, while still running a guerrilla PR campaign on Instagram and managing the growing calls for interviews about Zak's plight.

I had asked Zak to send me videos from Kabul to help me underline the urgency of the situation to lawmakers, journalists, and donors. The videos he had sent me from the post office as he sought the passports showed me that people in Kabul were clearly starting to panic as early as August 4. All I knew of Afghanistan was Helmand. I had never seen the more cosmopolitan areas like Kabul. Helmand had always seemed to hang by a thread, with even the provincial capital forever at back-and-forth risk of falling to Taliban control. Seeing evidence that the national capital was collapsing, seemingly under its own weight, was hard to comprehend. But I had to separate mine and Zak's fate from that of Afghanistan. I had a mission, and tears for anything else would have to wait till we had our own victory.

Andrea was understandably exhausted, and I had been awake for the better part of two weeks. I had not heard anything encouraging about the single remaining human resources verification letter that Zak needed, though I knew the retired officer was working it diligently, thanks to regular messages he sent me. I was getting close to desperation. No one could give me confidence that even if we got Zak on to the airfield at Hamid Karzai International Airport (HKIA), the State Department would allow him onto a plane without a passport for Deewa and the visa that Zak had been seeking for almost six years.

Any success I've had in life has come from a determination to overcome opposition regardless of the likelihood of success. But we needed to nail down the visa and deep inside I was beginning to lose faith that it would happen.

What St. John of the Cross called the "dark night of the soul" came for me on August 14, 2021. Andrea and the kids were long in bed. I was lost in reverie and somehow too tired to sleep. In Sangin, Zak had asked me what the phrase *Semper Fi* meant. The Marines said it all the time, both sincerely and sardonically.

I told him, "It is short for *Semper Fidelis*, which is Latin for 'Always Faithful.'"

He looked puzzled. He first asked, "What is Latin?"

I was well prepared for that one after a Catholic education, but we went on to talk about the meaning of "Always Faithful" and the immutable fact that it's got to be *always* to mean anything. Many people are intermittently faithful. Many people are faithful when conditions are favorable because when the sun is out and the goal is easy, it's easy to be faithful. That word "faithful" only means something if you attach "always" to it. That's an extraordinarily hard standard to meet. But that was Lance Corporal Arden Buenagua. That was Sergeant Matt Abbate. That was Corporal Tevan Nguyen.

Lance Corporal Arden Buenagua was the twelfth of thirteen combat engineers to suffer traumatic amputation or death in Sangin. My youngest son, born August 31, 2021, is named Marty Arden Schueman in honor of a man who knew the likely consequences and always moved to the front of the patrol anyway. Always Faithful is First Lieutenant Rob Kelly, carrying rockets on his back as an officer when he didn't have to and taking the hard route through the canal when he got killed, because Marine leaders lead by example. That's why my respect for Rob is so high. It's easy to lead by example for fifteen minutes when you have a belly full of food. It's hard to lead by example every minute. I know I've personally taken little selfish seconds or minutes for myself in the last fourteen years of service. Not Rob Kelly, not for a second. And not Zak for a moment of his service to the United States of America.

So when Zak told me, "Mr. Tom, it is time for *Semper Fi*," as we were discussing plans for getting him out before we both went on *The Rachel Maddow Show*, I damned well had to do my part to honor his commitment. That made it all the more galling to feel as if I was trying to keep hope alive for Zak as hope felt more and more like a lie, like the countless times over twenty years that politicians and generals said, "We're turning a corner in Afghanistan this year." I owed Zak "Always Faithful." It was time to stop climbing the mountain and move it.

On August 15, 2021, I drafted a supervisor letter myself. It said, in part, "I, Major Thomas K Schueman, was the Senior/Direct Supervisor of the below mentioned Afghan Local National Language and Cultural Advisor (Interpreter/Translator) . . . Zainullah Zaki's service as a linguist and cultural advisor was of extreme importance to my units' mission in Helmand Province. . . . As a result of Zainullah Zaki's service in support of the United States Armed Forces in the Afghanistan Theater of Combat Operations, Zainullah has suffered and continues to suffer threats to the life and property of himself and his family members by enemy forces and criminal elements."

I sent them to Zak and said, "Submit those immediately to your SIV package." It wasn't exactly what the program called for and it did not increase the length of his service as an interpreter, but I thought that surely, risking his life had to count for something, and in the deepening crisis, maybe my word would be enough.

After pushing SEND on the email to Zak, I reached back out to my Basic School classmate, Major Jared Lefaivre, on the ground in Kabul with the U.S. Air Force, and let him know I was sending Zak to the airport. I contacted a Marine captain on the ground at HKIA and confirmed that Zak could link up with him to pass through the Abbey Gate, which his company controlled, and then get an escort over to Jared at Ramp Eight.

The next message I sent was to Zak: "I'm going to try to get you a flight out today, pack your bags and get ready." I followed that with pictures of the gate where he would meet the Marine captain and a picture of the captain himself. My last call was to John Shattuck to tell him it was time to execute.

John had contacted Zak the day before to give him contact information for an escort named Siraj, who planned to drive Zak to the correct gate. At 11:45 A.M. Eastern on August 15, 2021, I told Zak it was time to get to HKIA. It was 10:15 P.M. in Kabul. John texted Zak to tell him it was time to call Siraj. He reminded Zak not to travel with any of his SIV paperwork, to bring only his family's *tazkera*—national identification cards—and passports.

I thought about the implication of that. It was as if someone called me late at night and said, "Get Andrea, Amelia, and Jack ready for someone none of you know to take you to Chicago O'Hare. Bring one bag each. Bring your passports and your Social Security cards. Forget your family pictures. Forget the kids' toys. Leave every item of your hard-won middle-class existence behind. And, oh yeah, you can never go home again."

Then I put that all away. We were headed outside the wire.

AS IN ALL complex operations, the breakdowns in the plan began almost immediately. As we had instructed him, Zak called his Kabul escort, Siraj, to tell him where he was and that he needed to go to the Abbey Gate. We knew the Abbey Gate was controlled by U.S. Marines under the command of the captain helping us. We also knew that desperate people, even those with no credible hope of flying, were beginning to flood the airport. Zak called Siraj, but rather than telling Zak when he would pick him up, Siraj said it was too dangerous to take him to the airport.

Okay. Plan change. The mission was now a foot movement.

All Zak needed to do was move on foot with his wife, four children, and baggage containing all of their possessions the four miles to HKIA, in the middle of the night, as the nation he had fought so hard for ceased to exist. He never questioned me. He just gathered his family and they started walking a few hours later.

I finally exhaled at 8:00 P.M. EST when Zak texted to say, "I just got there sir." It was 5:30 A.M. Afghan time and he and his family were at the passenger terminal on the south side of HKIA. The problem was we needed him at the military facilities where Jared was, on the north side of the airport. If HKIA is a rectangle, with the long edges at top and bottom, Zak was at the center of the bottom leg and I needed him at the center of the top leg.

"Zak, you need to get to the military side of the airport, to Ramp Eight." I had no idea he could not just enter the airport and cross once he got through the Abbey Gate. A voice message arrived on my phone.

"Sir, how can I get there? There's no taxis, my whole family . . . we need someone to escort, the situation is not good. We can't go alone." I closed my eyes and tried to think about a place to which I have never been.

"Okay, you need to go to the Abbey Gate. I'm working on the escort." I texted the Marine company commander at the gate. He sent me a picture of his finger pointing on a map to where Zak and his family needed to go. I sent it to Zak.

Nine hours before, the plan had begun as a ride to the airport for a link-up with Marines at the Abbey Gate, followed by a Marine escort across the airport to Ramp Eight, where Jared was. None of that had worked out, and now it set in that the original plan had made no sense given the actual geography of HKIA and the locations of the people involved. Instead, the plan changed to a four-mile walk with his family and everything he owned to the Abbey Gate, where I wanted Zak to ask the Marines to let him in and to meet Jared there. In execution, Zak had arrived at the wrong place after his four-mile walk with Deewa, four kids, and their baggage and I was telling him to leave and walk a mile to the Abbey Gate, ask for the company commander, and also have the Marines talk to me on the phone. If that half-assed plan failed, they would need to walk two more miles to the military side and call an Air Force pilot named Major Jared Lefaivre so he could come and get them, assuming he answered the phone.

We were not in a world of good options. We were in a world of reacting to the least bad option as it presented itself. Now Zak was just going to have to walk his family at least another mile around one of the most dangerous places in the world.

As I waited for him to lead his family to the Abbey Gate, I flashed back to the day in Sangin when Lance Corporal Arden Buenagua and Sergeant Jason Peto were killed and Gunnery Sergeant Chris Carlisle was wounded. I remembered looking at Zak as we sat in the overwatch position, unable to directly influence events on the ground, and saying to him how helpless I felt. I remembered the warmth of his hand gripping my shoulder, silently saying he was there for me and always would be. I hoped he could feel mine on his now.

Morning was breaking on August 16 as Zak and Deewa carried the kids and baggage to the Abbey Gate. The sound of jet engines screaming must have been terrifying for the children and the day was already headed to even hotter temperatures than normal. By the time he got near the military side of the airport, crowds had built to the point that people were pressed up against one another at the gates. Zak stopped to call Jared, as I had told him to. There was no response. Cell coverage around the airport was poor, likely due to electronic countermeasures employed against remotely controlled bombs.

The Marines were not responding, either. They had been thrown into a situation unlike anything for which most of them had prepared. Part of the Marine Corps mission is to be as prepared as possible for the worst contingencies imaginable and to flex to respond to the ones no one foresaw. But no one foresaw a flood of Afghans on the runway, or people clinging to the landing gear of Air Force C-17 cargo jets in a desperate bid to get out of Kabul before the Taliban consolidated their control. No one foresaw human beings plummeting from the sky as their fingers lost purchase on the cold metal skin of an airplane. But that's what I saw on the news at the same time I lost communications with Zak.

Staring out the window into the post-midnight dark, I was texting with no reply, promising food, promising water, trying to do anything I could to let Zak know we were not quitting. I called my Marine contacts. I called Jared Lefaivre. I called John Shattuck. No one had any idea where Zak was. I stared at walls and thumbed plaintive and unanswered messages to Zak, waiting to hear something; imagining Taliban checkpoints and children's bullet-torn bodies in Kabul.

Finally, after three hours I got a voice recording from Zak. He had never gone beyond the Abbey Gate. He walked his family back to the apartment after a hail of gunfire at the airport.

"I had a problem charging my phone but we're just returning back to our apartment because there was a gunshot, Taliban did it or American, I don't know who did. A lot of people got injured and died. So that's why we are returning back to our house. I already talked to Mr. John Shattuck. He says it's good, when the situation is controlled, then we will find out. . . . Find another way to get you out of there. Thanks, I appreciate and thanks for your time and your kind support. I appreciate you did a lot of things, you did a lot. Thanks. I appreciate that sir."

After fourteen hours, I had failed Zak once again. But he needed me positive and I had no intention of quitting. I messaged him at 2:15 A.M. on August 16: "Zak, you made the right decision bringing your family back to the apartment, I'm sorry that I could not get you on a flight today. I've worked all day making calls and we'll continue to work until we have you on a plane safe with your family. I'm thinking and praying constantly about your family."

He answered almost immediately: "I know sir. I know you did your best. We will see if the situation is better, then find out a better way to get out of the country, and thanks."

In the midst of growing anarchy, as I sat feet from where my children and pregnant wife slept peacefully, with his own family exhausted from walking more than eleven miles that day, Zak was still taking care of me. I walked into our bedroom and lay down next to Andrea to sleep for a few hours before school. I put my hand on her belly and felt my son move within her.

SITTING IN CLASS that morning, I felt like I was coming out of my skin. We had been in the local and national news at that point, so my professors were very understanding of what I was doing and why. Still, the fact that I was blatantly ignoring what was happening in class and either pounding on my laptop as I emailed with supporters and journalists or thumbing messages to Zak on my phone was an unintentional belligerence that felt very alien to me as a Marine. However, nothing is more central to Marine Corps leadership than "mission, men, myself," in that order. The mission required me to communicate constantly, and my man was in Kabul, not Newport, so I sat by the classroom doors and when National Public Radio called in the middle of class I walked out and I did a thirty-minute interview with them.

By the time I got home from class, Zak was almost overwhelmed. He was increasingly the subject of attention, and well-meaning people were flooding him with information that just did not, or could not, pan out. People were sending messages about routes through Uzbekistan and Tajikistan, which meant I had to spend energy telling him, "Don't leave with them and don't listen to them!" There were other people sending rosters around that supposedly would get submitted to the U.S. government, solving this increasingly wicked problem with names on a nice, neat flight manifest. Maybe there would be peanuts on the flight, too.

There is a concept in the military called "unity of command," which basically means there can only be one boss. That was me. Eventually I had to get very direct with Zak and say, "Don't do anything that anyone else tells you without asking me first." But I was not immune from fantasy, either. I still believed on August 16 that the U.S. Department of State was going to make good on the nation's promise and issue Zak an

SIV. I kept telling him, "Keep checking your email. They're going to say when it's okay to go to the airport. You might need to go soon."

In the depths of desperation, it was easy to want to believe in fairy tales. The Twenty-Fourth Marine Expeditionary Unit and the Eighty-Second Airborne Division were at HKIA. Ultimately, there would be six thousand American military members there. But the likelihood that anyone from those units was coming to rescue a guy whose service to America I had been unable to get anyone to acknowledge for five years seemed remote at best. I was rapidly accepting that we had to do this ourselves, with whatever team we had and could continue to build. Part of that team was an Afghan named Milad.

Milad was the result of another social media connection. A former British Army contractor in Kabul saw Zak's story on my Instagram and reached out to me. I don't know his story or what he was doing in Kabul, but Milad was one of his employees. He messaged me and said of Milad, "This guy's got balls of steel. He's your guy to do all this."

I had no time and no way to vet anything he said. After the first escort failed us and forced Zak and his family to walk more than ten miles, we needed someone like Milad because we were about to make the second push and things were only getting tighter at HKIA. I made contact with him and explained what I needed. He was all about it.

At 11:30 p.m. EST, I called Milad in Kabul, where it was 9:00 a.m. on August 17. He told me we were going to wait a few hours before we started the second effort. I started to object, then I thought about how angry it made me in combat when someone in a dry, air-conditioned operations center implied I was not seeing what I said I was seeing. I messaged Zak to tell him we would wait a few hours and to "make sure everyone gets a good breakfast and lunch. Make sure the babies get some naps and rest before you have to leave."

I looked at Andrea and said, "We have to do this. I can't fail here. These babies can't get left in Afghanistan without a father."

I got a message from the Marine company commander at the airport. It was time to move. The Marines were letting people through the gates, though precisely which gates would be open at any one time was subject to change. At 12:30 A.M. EST, now August 17, I called Zak and told him, "My friend at the airport says it's time for you and your family to go. Call Milad and see if he can take you to the airport right now. Go to the North Gate!" I gave him the names of the captain and gunnery sergeant from Echo Company, Second Battalion, First Marines. Echo, 2/1.

I followed with a picture of the North Gate and instructions for Milad on the route to get there.

At 2:30 A.M. in Newport, it was noon on August 17 in Afghanistan. Milad safely delivered Zak past Taliban checkpoints to HKIA, but Zak was disoriented on arrival. He didn't know which way to go. An hour later, I got a video from Zak. He was out of breath. "Sir, I'm at the gates, it's one o'clock, I am not inside yet."

I replied, telling him to tell the Marines at the gate that Major Schueman wanted to talk to them. An audio recording came through, announced with a ping. "Sir, the situation is not good! There's Marines here but they don't want to talk to anybody. They just say, 'Get out of here!' That's it. What can we do? Guide us. It is not good situation. Our kid is crying. It's a very danger for us." I was gutted.

It was 4:16 A.M. on August 17 when I said, "Give me ten more minutes. If I can't get someone to get you out, Milad will come to you."

As I typed and sent that, our messages crossed in the night. Zak's brought a picture. Zak was within speaking distance of a Marine as the Marines were closing the gate. I messaged for him, "Yell to the Marine, Major Tom Schueman wants to talk to him! Yell that the Gunny and the Captain said you're good." Nothing. Communications went silent. Then another video arrived at 4:35 A.M. EST. It was Zak's three-year-old daughter, Taqwa. She was sobbing in a way that was deeply familiar to me from my own daughter, Amelia. It was the whole-body cry of a toddler at the end of her endurance. And then gunfire. The video froze as it ended, framing her terrified face. I put my head in my hands and tried to breathe.

Nearing my own limit, I felt her cries in my heart. I was as close to genuine defeat as I have ever been. I dropped my head and leaned forward to rest it on my arms, crossed on the tabletop. Another message came through at 4:45: "I'm with Milad on the way home sir."

I sat up straight and exhaled. My thumbs went to work. "Stay safe, sorry it didn't work out. But we're going to get you out."

19

A Face in the Crowd

ZAK

I NEVER WANTED to live anywhere but Kunar Province. I served my nation and the American mission in Helmand to make Afghanistan a better place for my family to live, not because I wanted to leave. I was born into war and grew up in war and I wanted something better for my children, in the place where they were born. I wanted them to know their family and grow up where the bones of their ancestors lay. But I would not raise them in a place where my attempts to make a better life could get us all killed; I could not give up the freedoms I fought for. Since 2001, I had learned so much about what America means to the world. I wanted the freedoms and laws and security that Americans have, as part of my own life in Afghanistan. The Taliban and the corruption of our government made that impossible. Still, as I thought about leaving our home, I could not find it inside myself to rejoice.

The sound of a message arriving on my phone in the late hours of August 15 told me I had no more time to reflect on being forced to escape my own country. It was John Shattuck telling me to call Siraj and tell him it was time to go to the airport. I had never been to HKIA, so I was glad we would have a local guide to take us where we needed to go. He wished me good luck. I told Deewa to wake the children and dialed Siraj.

"*Bale?* Who is this?"

"Siraj, this is Zainullah. I am a friend of your friend. We are ready to go."

"Zainullah, it is a dangerous time now. It is not a good time to go to that place. We will try tomorrow." My heart sank.

"I cannot go tomorrow. My friends say it is time to come now. Please! I have money!"

"I cannot help you, Zainullah. *Mwafaq baasid*."

"Please! I do not need good luck! I need a ride for my family!"

He hung up.

I sent a message to Tom to tell him we would walk the four miles to the airport. I looked at my four children. Subhanullah was already up and helping his mother. Husna and Taqwa sat on the floor, leaning against each other. They could barely keep their tiny eyes open. Sadaf was asleep in Deewa's arms. I looked at our four bags. It was all we had in the world now. It had seemed very little. Now that we had to carry it, it seemed like a lot.

It is my wish that no other parent would ever know the fear of stepping into the dark streets of a dying city with their children. In the earliest hours of August 16, Deewa carried one-year-old Sadaf and held my son Subhanullah's hand. She had a bag on her back, as did my six-year-old son. I carried four-year-old Husna and three-year-old Taqwa in my arms; two bags on my back. We walked through silent streets. It felt like the stillness before a sandstorm in Helmand. The children were exhausted and slept in our arms. We arrived at the passenger terminal after two hours. I messaged Tom, only to find out we were in the wrong place. We needed to walk another mile to the Abbey Gate, where we hoped the Marines would let us through. If they did not, we would need to walk all the way around the airport.

As we walked toward the Abbey Gate, with blast walls rising thirteen feet above us on either side, the sun was upon us. It was still early morning and cool enough, but there was no shade. The sun reflected back up at us from the hard-packed dirt and concrete. At 7:00 A.M., the morning stillness was no more. Already there were crowds of scared people forming. The constant roar of their panicked shouting was even louder than the sound of jet engines from the airfield. As we walked, I remembered riding a U.S. Air Force plane from Bagram to Kandahar and how the sound of the engines overpowered my senses. I hoped we would be walking on one like it soon.

When we arrived at the Abbey Gate there was almost no way to get to it. There were too many people. I called the number that Tom gave me for the pilot named Jared. There was no answer. The Marines were not responding, either. I called to them, asking for the Marine company commander whose name Tom had given me. They did not respond to me; they just looked back at the growing crowd with empty faces. I saw some of them huddling to talk. Suddenly the atmosphere shifted. The Marines we could see now looked nervous and angry. They held their rifles differently.

I looked to my left, beyond the gate, and saw trucks approaching, carrying armed men with wild beards and long hair. The Taliban were at the airport. It was almost impossible for me to believe but they were part of the security there. They were pushing people back from the gate when, for the first time since Sangin, I heard gunfire and the sound of bullets cracking in the air over our heads. I ducked and held my family close, my back to the gun, my family held as close to my chest as I could get five people. A man next to us cried out and fell next to my daughter, Husna. From the way he lay, twisted and still, I could tell he would not get back up.

Looking at Deewa, her eyes told me what she was thinking. We could not stay at the airport with our children in these conditions. We moved with the crowd away from the Abbey Gate. The crowd was so close we almost had no choice. When we were far enough away, we started walking back to the apartment, reversing the way we came from the night before. The midday sun was bearing down on us. The bags on my back pulled at my shoulders. Taqwa, Husna, and Sadaf were crying. I was more worried by Subhanullah's silence. He just stared ahead and walked, bent slightly under his own bag. When we were back at the apartment, I let Tom and John know. The kids were exhausted. They lay down on the floor. Deewa and I did not have to put them down to nap; they were already asleep when I looked up from typing the message.

Afghanistan is different from America. In Afghanistan, men make the big decisions. But Deewa was done being quiet after seeing a man killed next to our children.

"Zainullah, we are not going back there! This is over! I have no passport! We have no visas! We will stay here in Kabul. I have family here. We will stay with them or we will go back to Kunar." I understood her, but I could not stay in Afghanistan.

"We cannot stay here," I said. "We cannot go to Kunar. I will be killed."

She was furious with me. "I have lost my family! I have lost my home! My children were almost shot this morning! We will go back to Kunar and we will figure it out! Our parents worked for Najibullah and survived the Taliban. We can survive the Taliban!"

I looked at her. My stomach twisted, as if my pancreatitis were coming back. I wanted her to be happy. I wanted my family to survive. I spoke quietly but firmly. "The Taliban hung Najibullah from a streetlight near here. I am sorry. This is my home, too. But we cannot stay here, Deewa. It is not possible. I will be killed. You will be forced to marry a Talib. We are going to America."

I lay down next to my children and tried to sleep in the heat of the Kabul afternoon.

Tom messaged me that night and said we had a new escort named Milad. Milad would take us back to HKIA for a second attempt the next day, August 17. John Shattuck assured me we had a detailed plan in place this time. Deewa was not happy with me, but she assented when I told her. I felt caught between angering my wife, disappointing Tom, and getting captured by the Taliban. All of it felt bad.

AT 10:00 A.M. on August 17, Tom told me to call Milad and get a ride to the North Gate. He again gave me the name of the Marine captain and a Marine gunnery sergeant who would be looking for me. He sent me a picture of the correct gate and the route for Milad to follow.

The streets of Kabul were busier than they had been the day before. I saw Taliban in the streets on motorcycles, but they did not stop us before we got to HKIA.

We pushed through the shouting and cursing of desperate people, again carrying the children and our bags. People held their passports in the air or held up signs written in English asking for help from the Marines. The Marines were doing what they could to pass water to people. They were the same age I was in Sangin. They looked like children to me now. I knew from living with my Marine brothers in Kilo Company that they wanted to help, but there was only so much to do. When we got to the North Gate, the Marines were telling Afghans to leave; no one was coming through the gate. We were so close, almost within touching distance of the Marine at the gate. I took a picture of him and sent it to Tom. He told me to tell the Marine that Major Schueman wanted to talk to him. I tried. I called to him but he looked through me. Tom told me to call out the name of the company commander or the gunnery sergeant. It did not matter. The Marines were shouting that the gate was closed. We had made it all the way there but could not go through. I called Milad. He said he was coming back for us.

As we were driving back to the apartment, Milad told me that the Taliban were looking for him, too. He said they had searched his father-in-law's home, looking for Milad, while he was driving us to the airport earlier. The Talibs killed his father-in-law and they would do the same to him. Milad said he had to stay on the move now, but he promised to keep his phone on and be there for us when we needed him.

I WOKE HOURS before my family on Wednesday, August 18, 2021. My kids had been awakened by nightmares during the night and now I could not sleep. I knew Tom planned to try again that day. I also knew that if we did not make it this time, I might not be able to convince

Deewa to try a fourth time. I had always found comfort in my prayers. Even in battle, I found time to speak to Allah. I faced toward Mecca and knelt to offer traditional prayers. But I also spoke to Allah directly: "Allah, the Most Merciful, if it is your will that I must die today, for my children to be safe, I will go happily. Please protect my family from the Taliban. Please get them to America."

I remained on the floor longer. I wanted to absorb the quiet. I knew I would need to draw on peace that day. Then I picked up my phone and sent a message to Tom: "I am still waiting on instructions."

Later that morning, at 9:15, Tom messaged me: "Zak my friend sent a message to me, he said 'come to the gate.' He said he sent you some pictures." Jared had sent me pictures of an unused gate at HKIA the day before. Now Tom was telling me, "Call Milad, you guys need to go. If Milad doesn't respond, you need to walk."

I knew better than to suggest walking to Deewa again. I called Milad. "*Bale?* Zainullah?" "*Bale, bale!* Tom says we must go to the airport."

Milad sounded tired. "I was there earlier. There's too much gunfire and too much crowd." I didn't know what to say. Milad was not a coward. He had not failed us. He was willing to help us again even though the Taliban killed his father-in-law. He just wanted to wait for a safer time. But I knew there were many Americans working to get my family out of Afghanistan. I did not know which way to go. I asked Milad to please wait and sent a message asking Tom what to do. Tom replied that I had to go because "the Taliban is letting people with SIV's through to the airport."

I felt as if someone had gripped my heart in their fist. I did not have an SIV. I had an application for an SIV. Now Tom was telling me that I had to show that to the Taliban. For years they said they would kill me. Three weeks before, they left a message on the gate of our home telling me to turn myself in. Now I was supposed to walk up to a Taliban checkpoint and show them that I was once a U.S. interpreter, was now trying to escape them, and believe they would help us get to the airport.

Tom had been talking to an Afghan interpreter living in the United

States. His name was Obid. Obid's English is better than mine. Because today was so important, Tom would explain the plan to Obid and Obid would call to explain it to me in Dari. Obid explained there was a secret gate that the U.S. government controlled. Major Jared Lefaivre talked to someone on the inside and they would let me through it. He promised to call Milad, too. He said Milad would make sure we got to the secret gate. I messaged back, "Got it, sir, I will do whatever I am told by you and your friends."

When Obid called me, I listened very carefully as he explained how to find the secret gate. It was not on any map or internet images from a satellite. Tom sent me map coordinates to the gate where Obid said Major Jared Lefaivre would be watching for me through a scope. He said for us to dress as we always did but to have one of the kids wear something bright or bring something bright to wave so Major Lefaivre could see us and have some soldiers grab us. He said we should expect it to happen in the early afternoon.

The plan seemed hard. I looked at the map coordinates that Tom sent me for the secret gate. I did not know how to read them. I sent a message to Tom: "Sir I don't know what to do with these coordinates. Give me the exact address and the person to get me and my family in."

I was tired and confused and frustrated. My children were exhausted. Deewa was sitting silently. I just wanted to know where to go and who to ask for so I could get my family to safety. But I called Milad and told him I understood about the airport but I needed him to come and get us. I heard him sigh loudly. "Very well, Zainullah. I am on the way. I am bringing my brother-in-law, Zaidullah. I need you to take him with you." I did not know what to say. That was not part of the plan but I could not say no to Milad after what he risked for us. I owed him for all he lost.

On the morning of August 18, the trip to HKIA was different than the two previous attempts. Kabul felt like a city waiting on a disaster. The banks and post office were closed. If people did not need to be on the street, they were sheltering in their homes. Sitting in the back, behind Milad and his cousin, I saw pickup trucks with Taliban standing in the back. I could not believe they were on the streets of Kabul. It had been almost exactly twenty years since they had run from Kabul in the face of the American arrival. Now they were back as victors and the Americans were doing their best to leave quickly. I tried not to make eye contact with them. I did not know how they would respond. I felt like they would know me.

The way to the airport was familiar now. As I expected, the traffic increased as we got closer. We passed families walking, as we had two days before. They looked as tired as I felt. People were more desperate to leave every day. Many were going to the airport with no means of getting on a plane; they just had no other hope. By then, all of us had heard what happened on August 16 when people rushed the airfield. One man's body was found on the rooftop where he landed after falling from the plane he clung to. I wondered what he thought that led him to grab on to an airplane taking flight. How desperate must he have been? That is the depth of hopelessness. It is people realizing that everything they believed in and worked for is gone.

Milad's voice brought me into the current moment.

"We are here, Zainullah. If I see you again, I hope it is somewhere other than Afghanistan! *Khoda hafiz*."

I placed my hand over my heart. "*Tashakor*, Milad, brother. May Allah protect you, too. You have done more than I can thank you for. I do hope to see you again, when all of us are safe and in a better place!"

I stepped from the car and turned to shake his hand. I looked in his eyes and he smiled at me. "Go now!" he said. "You have a plane to catch!"

I looked at my family in the backseat. There was nothing I would not endure to see them safe. "Let's go!" I said. Deewa gathered the children together. I picked up our bags and Taqwa. Subhanullah took my hand. Deewa took Husna and Sadaf. We plunged into the crowd and started walking once again.

The sound of the crowd again reminded me of my first plane ride. I could almost not think, but I had to, more than ever. The plan was confusing. I had to remember a lot of details to get to a secret gate. I did not know where it was. I checked my phone and saw I had a message from Tom. It said something different than what he had Obid tell me.

Now Tom was telling me to go to the East Gate and ask for a man named Samir and tell him a certain password. John Shattuck was giving me the same directions, too. I knew that our only chance was to do what Tom and John told me, but I was getting frustrated and confused. I did not know who had the best information, only that my family would pay for a mistake. I told them both, "There's a lot of people messaging me, and I don't know who to listen to." Both of them told me that Americans they knew got people through the East Gate, between the North Gate and the Abbey Gate, by asking for Samir and telling him the password. I agreed to do whatever they told me, but first we had to get past the Taliban checkpoint.

Walking toward the Taliban, I realized that the dream of a free Afghanistan was dead. They stood as if they belonged there. My country was just a memory now. We had to get to America to live the dream I had for my family. If that now meant dealing with the kind of people I had taunted over the radio in Sangin, I would do it. We walked closer still. I felt as if with each step the air was more charged with electricity. Some Taliban were hitting people in the crowd with the butts of their guns, trying to drive them away from the airport. I heard one shout in Pashto, "Why are you leaving, traitors? Don't go anywhere else! Just stay here in Afghanistan. Why are you leaving the country, infidels? The people you serve do not care about you!"

I stepped before a Taliban wearing *shalwar khameez* with a *destmal* draped over his shoulder. His hair flowed across his shoulders from under a turban and curled under his graying beard. He had filthy running shoes on his feet, but his AK-47 was clean. I was scared to speak to him. I held out my paperwork.

"Sir, I have my family with me. They are already ahead." He looked at my paperwork and I realized he could not read. I pointed at some people in the crowd. "They are my family; they are catching a plane. We are not going anywhere; my wife and I are just taking them to a plane and then going home ourselves."

He accepted my lie and waved us on. One more small victory over them felt good. But I was still confused about where to go. I told Tom, "I have passed the Taliban checkpoint. I'm close to the U.S. forces now but still crowds."

It was 12:45 P.M. We were at the Baron Hotel, near the Abbey Gate. I was confused about how to get to the East Gate to meet Samir and tell him the password. It was so hot. We were out of water and the children were fading as we fought against the powerful current in a river of desperate people.

The internet was failing. One of my phones was dying. I did not know how we would escape if we lost the phone connection. I told Tom. He said, "Okay, just get to an American and ask for Samir. Don't worry about the internet, get to an American, ask for Samir, and give them the password." I wanted to do what he said. I told him, "Got it sir. I don't know what to do. The gate you told me about is near this area."

I was not even sure which gate I was supposed to go to now, but I meant the secret gate, not the East Gate. I knew Tom was looking at his computer in America, trying to tell me how to get to somewhere he had never been, either. I sent him a pin of my location and he told me, "Move along the wall to the right. That's not the gate you're supposed to be at. Go to the right of the gate and keep walking."

I felt like we were on different planets, not just different countries, but I followed the wall toward the gate where Tom wanted us. Fighting through the crowd was like trying to carry my family through a jungle, pushing branches out of the way, until finally I could not move anymore.

We stopped and leaned against the cement of the blast wall. Messages kept coming from many people trying to help, faster than I could read or reply to them. Deewa slumped, holding Sadaf and Taqwa. Husna clung to my leg while Subhanullah sat atop our baggage. At 10:00 A.M., the sun had found us and there was no shade in which to escape. I sent a message to Tom to tell him I could not follow all the directions I was getting from so many people. Tom told me to ignore all the other messages and just listen to him. He created a group chat for just me, him, and Major Jared Lefaivre.

I sent them both a photo of my location and one of a tower on the airport wall near the Baron Hotel. Tom told me to wait there. Major Lefaivre knew where I was. Tom told him, "Jared, you got to go get them."

Major Lefaivre was running the combat operations center for his unit. He could not come. "I'm in the COC. It's fucking chaos. I don't have approval."

Tom insisted. I could tell he was speaking directly into his phone, forming texts. "Jared you got to go get them." And, to me, "Just stay there. Don't move until I tell you to. You see what Jared said about what he's wearing."

I looked at the picture of Jared in the text chain, an American major wearing camouflage, four flags arrayed behind him. All the Americans I had seen were in camouflage. How would that help me? I showed it to Deewa and tried to commit his face to my memory. Tom sent Major Lefaivre a picture of myself and my family that I had sent him and John Shattuck that morning. Our family sat down to wait at the edge of a crowd that was rapidly growing angrier and more desperate.

After sitting in one place for three hours, I told Tom I was afraid we would attract the Taliban before Major Lefaivre could come for us. The crowd was so big that people could move only slowly, pushing past one another. The Taliban were walking among us whipping people with electrical cords and fan belts to push them back from the gates.

"Get back, cowards! Get back, infidels! You who would betray your country! Now we will lead again!"

We watched as people jumped into a sewage ditch to get away from them. They stood up to their knees in human waste, calling up to the soldiers on the wall to let them in. I had wanted so much more for our nation than for Afghans to end up knee-deep in shit, begging for help from another country. But there was nothing I could do anymore. There were too many people with too many lost hopes. I looked away and closed my eyes.

I was disturbed from my thoughts by the sound of an incoming message from Jared. "I'm coming with a couple of shooters. Send me picture of what you see."

I sent them a picture of the Baron Hotel and the lookout tower on the wall in front of it. Jared looked at my picture and texted back, "OK, we are coming for you."

I texted "OK" but the transmission did not go through.

The cell coverage at the gates was failing again. He texted me: "Zak, are you getting this?" I said yes. Then again came a message from Major Lefaivre: "Zak, are you getting this?" I was receiving his texts, but mine were not transmitting. Tom repeated Major Lefaivre's text, but I could not tell him I understood, either.

I was watching the top of the wall, where Americans were moving behind coils of razor wire. I saw one of them step over the razor wire and climb up on the roof of a tower. The tower sat on top of a building that other Americans were standing upon. The lone American on top of the tower stood at the edge of its roof. Unlike the other Americans behind wire and cement, nothing stood between him and us but fifty feet of air. As he looked slowly across the throng, he leaned out so far, I feared he would fall into the crowd below. But he was not looking our way yet.

A Taliban fighter stood closer to the wall and was watching the crowd, an AK-47 hanging across his chest. He saw us looking up at the American on the tower roof. He looked in the same direction and reached out to alert his friends. I saw him look up at the American again. I could tell they made eye contact with each other. The Talib put his hand on the pistol grip of his AK. The American reached toward his hip but another American standing on a half wall below him reached out to grab his leg. He shook his head, "No," and the one on the tower moved his hand away from his hip. The Taliban below them made their hands into the shape of pistols and pointed them at the American as he carefully climbed down and stepped back over the wire. I kept my eyes focused on him. I could not see his face. I did not know if it was Major Jared Lefaivre, but I felt in my heart that this was the moment. Then I saw him lift his hand to his mouth and speak into a phone. A message arrived on my own in the group chat with Tom and Jared.

The text said, "Put your son on your shoulder."

I lifted Subhanullah over my head, over the shouting, sweating mass of people. He was wearing a bright blue shirt, as John Shattuck had told me to have him do early that morning. The American waved me closer. I grabbed my family and we fought through the crowd to the base of the wall, with our new friend Zainullah following.

As we pushed my countrymen aside, I was nodding at the man I now knew was Major Lefaivre. He shouted to me, "Do you have your family!?"

"Yes!" I yelled back to him. "And one more!"

Major Lefaivre turned to the three Americans next to him, all large men and heavily armed. I could not hear them all talking but from their gestures I could tell they were trying to decide how to pull seven people up and over the wall. They had begun to attract attention from more than the Taliban. Drowning people will reach for any hand they see. The crowd was beginning to move toward the base of the wall, below where they stood. The Americans finished conferring and began to move.

My eyes were locked on Major Lefaivre as he sent me a text: "We have to figure out how to get you to us at the door to your left."

I looked up from my phone as he discreetly gestured to his right. There was a barricaded door nearby, leading to a courtyard. People were already there, stretching their arms through the bars toward the Americans on the other side. One of the bigger Americans began throwing water bottles down from the wall and in the opposite direction to distract them. The people in the crowd were thirsty. They turned toward the water.

I pulled my family and Zaidullah toward the gate, pushing against the crowd. The Taliban were watching the frantic people grab water bottles. We reached the gate and I locked eyes with the man on the other side. He was large and soaked in sweat. As I watched him withdraw the bars securing the door, I saw his arms were covered in black and gray tattoos. I could barely distinguish where one ended and one began, but I saw an angel with her arms around the world and the words on his forearm, "That Others May Live."

He pulled back the last bar and yanked open the gate. I stepped to the side and physically touched my wife and each of my children, counting them as they rushed through the door, followed by Zaidullah and then myself. People started shouting and running toward us as the large tattooed man slammed the gate shut. One of his friends joined him in closing the bars to lock the gate shut. I looked out through the bars to where I and my family had been seconds before. People were already beginning to push against the door through which we had just passed. A man shouted to me in Pashto, his face crushed against the bars, "Help us, brother! Take my children with you!"

Then I felt the Americans grab me and push me forward so fast my feet were hardly touching the ground. They half-carried me to a car where they were already loading Zaidullah and my family. Before I got in, I looked back at the gate. The man was lost in the crowd on the other side.

AS WE DROVE, I saw I had service. I texted Tom at 11:50 A.M., "I'm so happy, inside."

He answered immediately, "So happy you're safe. I'll see you in America. I love you, brother."

Major Lefaivre took us to an air-conditioned trailer. After hours in the crowd, under the sun, it felt so cold. But he had water and chocolate and diapers for my children. While Deewa and I looked after them, a British man who was there shook the major's hand and said to him, "Good on you! It's a needle in a stack of needles looking for a guy in that madness, mate!"

Major Lefaivre looked at me and gestured at Zaidullah as he asked, "Who the hell is this guy?"

I explained about Milad asking me to take care of his brother-in-law after the Taliban murdered his father. Zaidullah told us then that he had a wife and two children still in Kabul. Major Lefaivre looked at him for a moment. The major looked tired. He took a breath and let it out before he asked Zaidullah, "Can you get them here?"

Zaidullah called Milad. When he pushed END on his phone, he smiled and said they would be there in an hour. Major Lefaivre looked at the large men standing against the wall of the room in which we were all sitting. I saw they had the letters "PJ" on patches on their sleeves. Major Lefaivre smiled and said, "I think we're going to have a lot more work."

I was so tired I was almost unable to move. I just stared into space and held my children. As we rested on the floor of the trailer, Major Lefaivre called the State Department consular services people at the airfield. He had them on speaker so I could listen.

"Hypothetically, if we grabbed some people, how do we get them out?"

There was a moment of silence from the other end. "Hypothetically, you shouldn't have done that, but if you hypothetically did, you could hypothetically use your rank to walk him up to the plane and let the Department of State sort it out down the road in Qatar."

After a few hours, we left Zaidullah behind to wait on his family and Major Lefaivre escorted us to the passenger terminal. We got my family seated and I went with Major Lefaivre to talk with the Marines registering people for a plane. They asked who we were.

Major Lefaivre told them, "Congressman Moulton has been texting me directly about Afghans he is worried about. He was a Marine. So was I." He pointed at me and said, "This guy fought with 3/5 Marines in Sangin. He and his family are passengers of high value. Plus, remember what Admiral [Peter] Vasely said: no more C-17s flying half-empty, he wants asses in seats!"

The Marine paused for a moment, then smiled and got us wristbands. He told me to get my family and put us in line for a plane. I called my family in Kunar to let them know we were going to fly to Qatar and that's all we knew.

I sent Tom a voice message: "Sir, we just got done with all the process, waiting in the terminal for our flight. If there's anything I will let you know. Sir, thanks for your hard work and support."

Major Lefaivre stayed and watched us until we started walking toward the plane. As we neared the plane, I turned to wave at him. I had to strain to hear him shout through hands cupped around his mouth, "Hey! Zak! *Semper Fi!*"

20

Return to Base

TOM

"SIR, THE SITUATION is not good. There's Marines here but they don't want to talk to anybody. They just say, 'Get out of here,' that's it. What can we do? Guide us. It is not good situation. Our kid is crying. It's a very danger for us." That was the voice message I got at 3:30 A.M. on August 17, 2021, as I was failing Zak for the second time.

That recording is the closest thing to a complaint I've ever heard from Zak despite two failed escape attempts and success on a third that seemed to come more from providence than our attempted planning and coordination. Through days that included his family's fruitless ten-mile walk to and from HKIA with everything they had in the world, a man falling dead from Taliban gunfire next to his huddled children, and unfulfilled promises that were made before he was even old enough to go to war on our behalf, Zak was stoic and resilient.

Zak's composure through circumstances unimaginable to most Americans contrasted starkly with my anxiety during the forty minutes of silence that followed a photo of the tower on which I knew Jared Lefaivre stood. The picture came at 1:39 A.M. EST on August 18. I knew we were close, but we had been so close the day before. I was torn between my need to know, my need to control the action, and my understanding that sometimes you have to shut up and let the people on the ground work it out.

And then, at 2:20 A.M. EST, from Zak, "I'm so happy, inside."

I replied, "So happy you're safe. I'll see you in America."

I had been awake for most of seventy-two hours. Suddenly, sitting at my dining room table felt like walking back inside friendly lines after an insane experience in combat. I was devoid of energy. I thought of the one cigarette I have ever had, a Camel Blue I smoked with Zak after we survived a run through IEDs outside FOB Inkerman. I wished I had a second one now.

I needed to sleep. But I also needed to reflect and thank God. So I sat and waited and prayed and smiled as he sent me pictures of his children, safe with Marines inside the airport.

I sent Zak a message that said, in part, ". . . I can't wait to have you in America. We're going to have a meal, your kids with me and my kids."

At 6:30 EST, Zak and his family were on a plane to Qatar. I posted a picture of Zak and his family with their Air Force rescuers to @kill .z0n3 with the caption "Wheels up" and finally lay down to sleep for a few hours. I awoke a few hours later to my children, Amelia and Jack, crawling on me. The sun was streaming through the bedroom window since we did not yet have blinds. The kids' excited chatter brought Andrea in to grab them and let me sleep. I looked at her, framed in the door, her hand on her pregnant belly making our third child's imminence so apparent, and stretched out my arm to her. We all lay together. One young family, inextricably tied to another on the opposite side of the world. I thanked God for blessing us all.

We still had no visa and there were still about 6,500 miles between Zak and American soil, but I assumed I would see him in a few days and help him get to San Antonio, Texas, to unite with his cousins already living in the United States, all former interpreters. I had no idea of the new journey we were commencing.

Zak arrived in Qatar on August 19, 2021. The children were exhausted. Zak sent me wrenching photos of Subhanullah, dehydrated, exhausted, and staring into space on a green military-issue cot with an IV in his arm. He also reached out to John Shattuck from the reception center: "Sir we are in Qatar now. It is not much better place. My son is sick now. And we are in a bad and dirty place waiting for our visa process. I don't know how long it will take. . . . He start vomiting and the medics are take care of them. We all good."

Zak, Deewa, and the kids were each assigned a six-digit tracking number and moved into a single room in a prefab building at Al Udeid Air Base to await onward movement. Deewa still had no passport, but the sense of overwhelming relief and appreciation Zak felt remained palpable when we talked.

The plan to move people to third countries appeared to have been overwhelmed by the number of people to be moved and the haphazard manner in which many of them did. President Biden referenced 120,000 evacuees in a speech on August 31. Forty thousand Afghans had arrived in Washington, D.C., and Philadelphia by September 3. But the official flights, with people like Zak who were manifested and accounted for by the U.S. Departments of Defense and State, were also accompanied by private charters paid for by well-intentioned people around the world. Those planes carried hundreds of Afghans who arrived with no accompanying clarity about their status or paperwork. Some of them entered a stateless purgatory. Some of them never should have flown. All of it conspired to slow Zak and his family as much as the absence of a passport for Deewa. "The system" for resettling Afghans was overwhelmed.

In the midst of those frustrations came good news. The retired officer who had been working to find verification of Zak's employment by the U.S. government had pulled every string and worked every contact he had among people who were nameless and faceless by design and identified Zak's employer. Someone in his network knew the American man standing with Zak in the picture I had given him. That picture led to a single old work order connecting Zak to the company. Those two items were the threads the retired officer and volunteer needed to unravel the curtain blocking our ability to find who we needed to talk to. On August 20, the chief executive officer of the company, himself an Army vet reached through Marine and Navy connections, happily sent a letter attesting to his personal management of Zak from January 2012 to December 2013. We had the required amount of time in support of the U.S. government for Zak to finalize his SIV application. After almost six years, we had reached step two of the fourteen-step process.

Finally, on September 9, after three weeks in a single room in Qatar, Zak and his five family members were flown to Ramstein Air Base in Germany. There they lived in a tent camp for five weeks. Along with thousands of other evacuees, they underwent biometric screening and interviews to ensure they were not a threat to the United States. They received medical evaluations to ensure it was safe to move them on to the States, both for their sake and the citizens of the United States. Zak and his family had left Kabul in August, when the temperatures were in the 90s. Now they were in Ramstein, where fall is rainy and temperatures are in the 40s. Being cold is a small complaint compared to being dead. But as a father I knew how hard it would be to look at my children and not be able to explain when we would be somewhere they would be comfortable again.

Zak and his family finally flew to America on October 15, 2021. They spent two days in Philadelphia before being assigned to a resettlement center at Fort Pickett, Virginia. There his family lived in a single square half-mile area from which they were not allowed to leave. They spent more than three months in barracks designed as temporary lodging for National Guard troops mobilizing for deployment, not extended living for large families who separated their living spaces with sheets and army-green wall lockers. Zak and his family lived in a space smaller than some American master bathrooms.

For all the difficulties, the resettlement centers were staffed by truly caring people doing their best to see future Americans established as well as possible. There were a host of resettlement organizations: religious, charitable, and human rights focused (or some combination thereof), working to lay the foundations for people's new lives as possible. But as in all things governmental, the resettlement process is like a pipe through which only so much can pass at any time. John Wiles, a resettlement worker, was particularly generous with his time in fielding questions about the mechanism for moving Zak to his final destination and translating the reasons for delays. A few conversations with him helped reframe the plan.

Any complex system looks simple from the outside. When you pull back the cover and look, some of what appears byzantine from the outside becomes clear. Some of it still seems byzantine. There were requirements to be accomplished at Fort Pickett to prepare new arrivals for life in America. There were bureaucratic hoops to jump through as well. Those things were easily intuitive for a guy with fourteen years in the Marine Corps. What I was not aware of that was happening as early as August 31 was the search for square resettlement holes for square pegs and round resettlement holes for round ones.

If you're a single male, speak English, and you're flexible on destination? Easy, you're going to Phoenix, or Anchorage, or Atlanta. But if you're a family of six, with a spouse who speaks no English, and four kids under six years of age? Requirements start to expand and destination possibilities contract. You need an elementary school that is ready to help kids who are learning English. You need a mosque, preferably in walking distance. You need a low-cost or Section 8 apartment or a charitable group that will pay for another one. You need grocery shopping that is remotely recognizable to you and your wife, who can't drive and whose lack of English makes public transportation a challenge. You need a resettlement organization that has not hit its caseload cap and is in a city and a state where both are receptive to Afghan immigration. Not all were.

Deeply dedicated people had been doing the best they could with what they had on behalf of thousands of people like Zak since August 15, 2021, when the Taliban rolled into Kabul and the exodus began in earnest. Before I began to comprehend the convolutions of humanitarian resettlement, I had planned to drive a truck to Fort Pickett and drive Zak to San Antonio in December to live with his cousins. As I began to grasp the scope of the issues, I was reluctant to say much even when it became clear that our plans for a December move were not going to happen. Still, it was frustrating. I had not seen Zak in person since FOB Inkerman in 2011. I wanted to see him just to verify with my own eyes that he was well. Zak was happy with my plan and ready to go to San Antonio, where he had family who wanted to receive him. But to stay inside the benefits system built to assist resettled Afghans, he had to follow the prescribed steps mandating that he, Deewa, and the kids go where they were offered an opening appropriate to their needs. They were free to move on to wherever they wanted to go once that was complete, but until then, they were along for the ride.

At the end of January, things broke loose, like ice cracking and flowing down a river. The opening was in Minneapolis if he wanted to get out of Fort Pickett. Zak said yes and his family flew there on January 26. It was at least a step closer to where he wanted to be.

By February 19, I could not wait any longer to see Zak. I hopped on a Friday evening flight. Things did not start well. I was delayed out of Providence, Rhode Island. Fortunately, my Chicago flight was delayed just enough that I could make the connection by sprinting for ten minutes, carrying bags heavy with presents for Zak, Deewa, and the kids. I sat and poured sweat in economy seating for half the flight to Minneapolis. All that was forgotten the next morning when Zak and I connected in the hotel.

TIME IS THE great equalizer. Zak and I are both fathers and husbands in our thirties now. We are men who once stood together in one of the most dangerous places on earth. We are brothers. Riding down the elevator from my hotel room, I realized that Zak is something even more than a brother now, especially in an era when the word gets thrown around as a collegial shorthand. Brotherhood is a status born of blood or shared experience, and that is no small matter. But for me, Zak occupies a status in which one can only arrive by their active choices, by realizing a person is someone for whom you would truly move heaven and earth and that he would do the same for you.

Despite that elemental connection, as I met Zak on a Saturday morning in a Minneapolis hotel lobby, I was nervous, excited, and so, so thankful. As I saw him for the first time in a decade, I felt, heard, and smelled the fullness of the experiences that led to the active choice that leaves us more than brothers: fear, angst, and joy; explosions, laughter, and screams; sweat, smoke, and fresh-turned earth. I felt a deep love and gratitude to thousands of people, to God, to Zak and his family.

Zak and I went to my room and talked for hours about where we had been together, the years between, and what life might look like now going forward. Zak was his usual gracious and appreciative self. "Minnesota is a nice place," he said, "a good place with kind people who show us kindness." He continued: "We are getting good help from the resettlement people at Church World Service and they will help us when we get to San Antonio. The kids are in school and happy to go every morning. They have pencils and everything, learning the ABCs and one-two-threes." He laughed as he said, "But I'm pretty cold."

I thought a minute. "So all we need is your Social Security card and we can move you to San Antonio, right?"

He looked puzzled. "Mr. Tom, I have my Social Security card." I am nothing if not decisive.

"Well then, let's go to Texas." We booked the tickets for the next day right there and then.

Just after midnight on February 22, 2022, seven passengers, eight bags, and four car seats arrived in San Antonio. Two of Zak's cousins met us at the airport, along with Travis Haggerty, a Marine from First Platoon, Kilo Company, 3/5. The hugging, gracious welcomes and loving protestations over who would be allowed to carry whose bags bled into the cousins being aghast at the notion that I was headed to my airport hotel to sleep before an early flight back to Providence. At 12:30 A.M., while loading baggage in the arrival lane, the cousins would not allow me to get in the car until I had committed to come to their house for a party and dinner with them. My hotel was five minutes away and it was after midnight. I desperately wanted to sleep before flying the next morning. Despite my pride in my ability to shape a situation into conformity with my will, thirty minutes later I arrived at his cousin's house to a joyful gathering of children, moms, and dads.

Zak and I sat beside one another in a traditional lounging room, drinking endless small glasses of blistering hot chai with the cousins. Then food arrived, served upon huge platters as we sat Afghan-style on the floor. It was wonderful and relentless: *Kabuli pulao*, kebab, naan, *bolani*, and food I didn't know, but which smelled and tasted delicious. When I demurred, the only answer was "No, eat! You eat!"

I looked over at Zak and asked him, "When's the last time you had a meal like this?"

He looked back at me, his face a mixture of both sadness and joy, and said, "August seventeenth. The day before we left Kabul."

The stories went late into the night. I told the story of my airport sprint through Chicago O'Hare with heavy baggage that left me pouring sweat on the airplane. Zak looked at me with a gleam in his eye as he said, "That sounds horrible, like walking ten miles with your bags and kids through streets where the Taliban are shooting at you." It was a joke, but like when he assured me that my hotel tap water was fine, that I did not need to go twenty-two floors down to a filtered water bottle filler, I realized the stark reality he'd been living compared to our own in America.

Freed by the relief of being in the place where he'd been trying to get since August, Zak finally shared with me the weight he'd felt in getting from Kunar to Kabul and beyond: lack of sleep, the pressures on a father of a family in mortal danger, Taliban threats, the uncertainty of one moment to the next, the danger in which beloved family members still find themselves, to name a few.

Gathered around us were men who understood those pressures implicitly, men who served in combat with U.S. and Afghan security forces. All of us had hoped for a better Afghanistan. All of us had sacrificed for a dream now gone. There would be time, lots of it, for those present and millions of others to turn over the big question of whether it had all been worth it. There would be time for acrimony, recriminations, and debate over whether the lives lost and the money spent had been wasted, or whether there is value and meaning in sacrificing oneself for closely held ideals. The Kabul Airlift, and the war itself, taught us that among the venality, inhumanity, and sadness of war, there exist in equal measure kindness, self-sacrifice, and a willingness to move mountains that seem insurmountable. It taught us that love cannot conquer all, but it can conquer some, and that's a fact that should never be lost in the darkest moments of our collective nights.

The reckoning was there and waiting for us in the future. It still is, but in that moment, it couldn't have felt more distant or less relevant. For now, as I sat and looked around the room, as I listened to Deewa laughing downstairs, so happy to be back with family, and heard the kids scream with delight as they leapt from couch to couch in the basement, I realized everyone here was now part of a broader dream, an experiment that began in 1776.

No one who didn't believe in the fundamental values of freedom and liberty that our nation holds as self-evident truths would risk the things Zak did. Just like Zak's saga, our collective journey hasn't been easy, or pretty, or safe, or fair. There have been times that we have fallen so far short of the mark as a nation, times we've been so grossly unjust to our own citizens, as to be almost unrecognizable to ourselves. But gathered with a group of future Americans, eating traditional Afghan food in a traditional Afghan way in a town home in San Antonio, Texas, I could not help but be flooded with all the hope and joy I had suppressed over five years. Mission accomplished. I was ready to return to base.

As I stood to leave, I remembered a final gift for Zak. I had believed we would part at the airport. The pocket copy of the Constitution of the United States I brought to give him there was still in my own pocket. I pulled it out and handed it to him. He opened and paged through it then asked, "Can I keep this?"

I pulled him close for a hug and nodded as I looked him in the eyes. "Yeah, it's yours now, brother."

EPILOGUE

TOM

IN LATE FEBRUARY 2022, I was sitting in the Naval War College library studying when my phone buzzed. I looked down at the screen. Seeing the name "Jerome Greco," I answered immediately as I got up and headed outside the library. "Sir! Darkhorse 6! How are you?"

After I came home from Afghanistan in 2011, I was moved from Kilo Company to Lima Company where I served as then–Captain Greco's executive officer and watched him give a master class in Marine Corps leadership while in command. As such, I'd cheered when, whether by providence or poetry, he was selected to command 3/5 as a lieutenant colonel in January 2022. When I talked to him on the phone a month later, he sounded happy.

"Hey Tom! I am well, how are you?"

"Great! I am finishing things up here in Newport. Andrea and the kids are happy and healthy. I am ready to get some."

Typically, a major coming out of the War College can expect to be sent back to a battalion for duty as an operations officer or executive officer, the third and second most senior members of a battalion respectively. As I neared the end of my time at the Naval War College, I had been sending out feelers about jobs to Marines I knew at Camp Pendleton, CA, and Camp Lejeune, NC, in an effort to control my own destiny. It wasn't lost on me that Darkhorse needed an XO.

"Well, Tom . . . that's what I am calling about. I am putting the band back together. You ready to come home?"

"Are you serious, Sir?"

"Yeah man! I need a battalion XO and you're my first choice. Come 'get some!'"

"Roger Sir, I'm honored to be asked. You know there's nowhere I'd rather go than 3/5. I already prepped Andrea for California if I could get assigned there. What do I need to do to make it happen?"

"Let me get the ball rolling, I just wanted to make sure you were in."

"Roger, Sir. This is great news."

Darkhorse. 3/5. Back to the beginning. I was excited to get back to the infantry, to again be a part of the blood coursing through the veins of the Corps. For a student of poetry, there is a beautiful symmetry to the journey. But it's a loaded one, too.

With the U.S. withdrawal from Afghanistan half a year behind us, many questions that it had raised anew about my first tour at 3/5 remained unanswered. How do you reconcile a twenty-year death spiral? How do you eulogize a dream? I don't know. As a Marine, I am just an instrument of the state in an all-volunteer military. The essential question comes of my time in Afghanistan, again and again, "Was it worth it?" But that question is not even mine to answer. The state demanded someone go. I raised my hand and said yes.

Whether it was worth it is a question for the American people, for though some of the effects of war are borne only by a very few, the costs are shared across the whole of the nation. In *Marine Corps Doctrinal Publication 1: Warfighting*, the Marine Corps asserts, "War is among the greatest horrors known to humanity; it should never be romanticized. The means of war is force, applied in the form of organized violence. It is through the use of violence, or the credible threat of violence, that we compel our enemy to do our will. Violence is an essential element of war, and its immediate result is bloodshed, destruction, and suffering." Many of those costs are invisible to a public far removed from service. As a nation, we might be more reluctant to pay them if they were more manifest.

Alfred, Lord Tennyson captured the truth of the military profession in 1854 with, "The Charge of the Light Brigade,"

> *Was there a man dismayed?*
> *Not though the soldier knew*

Someone had blundered.
Theirs not to make reply,
Theirs not to reason why,
Theirs but to do and die.
Into the valley of Death
Rode the six hundred.

Darkhorse lived Tennyson's truth one hundred and fifty-six years later. Sergeant Matt Abbate rendered it to its essence in late 2010 when he wrote, "Someone Must Walk the Point" on a wall at PB Fires. Despite that century and a half, the similarities of the experiences are stunning and offer some insight into the immutable nature of war.

In examining the worthiness of it all, I don't think there is any helpful truth in the numbers, they simply are what they are. Measuring them against concepts made ethereal by the Taliban's assumption of power, things like "democracy" or "freedom" or "prosperity," seem almost disrespectful to the very present losses in blood and treasure. Nonetheless, for posterity if nothing else, I must note that 980,000 Americans walked the point in Afghanistan during our almost twenty years there, saying by our presence that we did indeed consider Afghanistan worth it. For some, our very lives were worth it: 2,445 Americans died in Afghanistan. For many more, our welfare was the cost: 20,740 Americans were wounded, thousands of those grievously. Those numbers say nothing about the invisible wounds almost ubiquitous amongst those who fought the post-9/11 wars, such as traumatic brain injury and post-traumatic stress disorder. They are much of the reason Patrol Base Abbate exists. Afghan numbers are tragically imprecise, but at least 66,000 members of the Afghan National Security Forces died during the war. Some 47,000 Afghan civilian deaths are accounted for, a number that seems frankly low. Millions of Afghans are certainly displaced, even more so since August 2021.

But if I must answer the question for myself alone, yes, it was worth it. I volunteered to walk the point along with millions of other Americans. America sent me to Sangin to do so. I was granted the opportunity to see young Americans at their finest, to see men barely out of boyhood accept the reality of their mortality and yet run to the sound of the guns. I was gifted the chance to meet a brother from the other side of the world and welcome him into my family as he welcomed me into his own. Alongside a platoon of Marines, we met the demands of duty in one seven-month period. Over time, what began as a period of national service has become my profession.

After deploying to Afghanistan a second time, I commanded a company of Marines before being given the opportunity to contemplate the broader reality of warriors and warfare at the United States Naval Academy, an institution in which they are made. I spent another year studying and contemplating my profession at the United States Naval War College, a time that overlapped both the evacuation of Kabul and the writing of this book. During those four years of focused professional study, I had to reconcile my love for the heat of combat and my abhorrence of its waste. Accordingly, I believe we must use our warriors in pursuit of peace rather than in an endless cycle of war without end. These are ideas I believe wholly in keeping with my faith, my philosophy, and my commissioning oath.

I would have of course preferred to "win" in Afghanistan. Losing is anathema to me. But as I wrote in August of 2021, anyone surprised by the end in Afghanistan just failed to read the tea leaves. And there was beauty in the darkest moments. I lost a war but gained a brother, a sister, nieces, and nephews. Servicemembers and civilians alike moved heaven and earth to rescue Zainullah "Zak" Zaki and his family. Without people like Major Jared Lefaivre and countless unnamed people all over the world, Zak would almost certainly have met the same fate as the 490 people the *New York Times* has confirmed as murdered or forcibly missing as a result of Taliban action as of this writing. To allow that to happen to Zak would be a violation of our fundamental principles as Americans. It would be a failure of the motto, "Always Faithful."

One of my greatest honors has been, and remains, calling myself a United States Marine. Now, I prepare to take my ideals back to service in the Marine Corps operating forces.

As a teacher, a student, and a Marine, when I reflect on this journey, I wonder what lessons we've learned. C. S. Lewis wrote: "None can give to another what he does not possess himself. No generation can bequeath to its successor what it has not got. . . . if we are skeptical, we shall teach only skepticism to our pupils, if fools only folly, if vulgar only vulgarity, if saints sanctity, if heroes heroism." The last two decades have left us all feeling a little skeptical and foolish. It also produced a bounty of heroes. I hope that through a collective remembrance of those who perished in the fields of folly we bequeath our successors a peace worthy of their sacrifice.

ZAK

MUCH IN OUR lives is different now, but I still rise early to pray and offer thanks to Allah. On Fridays I go to the Islamic Center to pray, as my father did. I enjoy the peace of prayer before I go to work. Kneeling, I say three times, *"Allahomma ma asbaha bi min ni'matin aob bi'ahadin min khalkhika faminka wahdak. La sharika lak. Falakal hamdu walakash shukr."* "Oh Allah! Whatever blessings I or any of Your creatures rose up with, is only from You. You have no partner, so all grace and thanks are due to You."

Three times again, *"Ya rabbi lakal hamdu kama yanbaghi lijalali wajhika wa'azeem sultanik."* "Oh my Lord! All grace is due to You, which is befitting to Your glorious presence and Your great sovereignty."

Three times more, *"Allahomma anta rabbi la ilaha illa'ant. Khalakhtani wa'ana abdok w'ana ala ahdika wawa'dika mastata't. A'ootho bika min sharri ma sana't. Aboo' laka bini matika 'alayya wa'boo' bithanbi faghfirli fa'innaho la yaghfroth thonooba illa'ant."* "Oh Allah! You are my Lord. There is no deity but You. You created me and I am your slave-servant. I am trying my best to keep my oath of faith to You, and to seek to live in the hope of Your promise. I seek refuge in You from my greatest evil deeds. I acknowledge Your blessings upon me, and I acknowledge my sins. So forgive me, for none but You can forgive sins."

When I am done, I kiss my sleeping children. I kiss my wife as she prepares food for them when they wake. Then I drive to work. I have time to think as I drive. I cannot forget the things that have happened to me and my family. I ask Allah for help sorting my memories. But I cannot forget the things that people did to help us either. I would be dead if I were still in Afghanistan. My wife and children would be in mourning or in the ground beside me. Instead, people risked themselves to save us and now we are surrounded by friends and family in America.

Everywhere we have gone in America, we find good people. Virginia, Minnesota, Texas, all have welcomed us. Life is good in San Antonio. Through the kindness of many people, we have all the things we need to make a life here. We spend time in prayer at the Islamic Center. Nearby we have halal groceries that sell Afghan food. My children love the parks and have so much fun watching the ducks. We go to a nearby lake and I tell them about the Kunar River flowing cold from the thawing Hindu Kush. We have a car I bought myself, my first. I have a job cleaning up after building demolition. It is honest work that supports my family, and I am proud to have it. I will do it for as long as they will have me or until another opportunity comes, like anyone else in America. I am very proud that my son will begin school in the fall, prouder still that my girls will follow in the coming years. We are blessed to be here. But I never wanted to leave my home in Afghanistan. I doubt I will ever see it again. If I do, I do not believe I will survive the journey.

Afghanistan is the place of my birth, the place where my family has always lived. My grandparents and parents and siblings are still there. They are not well. It hurts to be unable to help them. I wanted so many things for our country. Maybe I hoped for more than the country could accomplish. Maybe we were too fractured to ever truly call ourselves a nation. But for a moment, the dreams of democracy, prosperity, and freedom for my family and my fellow Afghans, for all religions and all tribes, seemed within reach.

With the departure of the Taliban and the arrival of a world that had ignored us for years, except to extract what they could from our poor country, anything seemed possible if we just worked hard enough and were willing to risk enough. I was willing to risk my life. Thousands of my fellow Afghans lost theirs. I worked for my people and country in pursuit of a better future for all of us. But I also worked with and for the Americans, because they came to help us build a better Afghanistan. I would do it again. I would still sacrifice myself for my country, for my people, and for America, the country that saved and welcomed me and my family. Too many good people died for the dream for me to ignore the obligation to pay it back.

The Taliban were not our only problem. The corruption in our own national government was a chain around our collective neck. Leaders must earn their positions by deserving the trust of the people. A government with anything less will never survive and the governed will never thrive. Too many of our leaders bought their way into power as a means of enriching themselves. Too many warlords held on to power by force, placing themselves ahead of the people they were supposed to protect. It was a frail system that fell when the foreign support holding it up ceased to exist.

For now, the Taliban own Afghanistan again. The Afghan people are again living a nightmare. People I love who are still there are suffering. They are starving, with no food and no money with which to buy it. Modern medical care is already only a memory. Security and freedom are a dream. When they seized the nation, the Taliban claimed that former members of the security forces and people who worked with the Americans would be safe in a Taliban regime. It was a lie then and it is a lie now. They claim to be a new Taliban, different from the gunmen who swept out of Kandahar in 1996. They claim to be ready to lead the "Islamic Emirate of Afghanistan" in the modern world. They are not. A savage with a Twitter account is still a savage. You would not allow a viper into your home and expect anything other than misery. The executions and torture have already begun. I knew that was coming, the Taliban told me themselves.

On August 18, 2021, through the kindness, bravery, and dedication of Americans and Afghans alike, I left behind almost everything I have known and loved to escape the Taliban with my wife and children. Now we are safe and happy and all we want from life is to stay that way. But on March 22, 2022, my application for a Special Immigrant Visa was denied by the chief of mission on stationery from an embassy that no longer exists. There was no phone number, no address to write, just another email address no one will respond from. I get only one appeal. I hope and pray to be allowed to remain in America so my wife, Deewa, and I can raise our children to honor the freedoms that make America special in the world.

To be denied a chance to even apply for an SIV after combat, six years of applications, death threats, and my family's flight from our country was heartbreaking. I immediately felt hopeless. But then I remembered all we have overcome. I remembered *Semper Fidelis*. I again turned to Major Tom Schueman. I spoke to other American friends I have made. I talked to my immigration attorney. The Marines of 3/5 never quit. I never quit. I will not start now.

If I am again denied the chance to apply for an SIV, I will be required to show past persecution and likely future persecution as part of asking for asylum in the United States. I feel that is obvious. If I return to Afghanistan, I will be killed. If my wife and children are allowed to survive me, they will be punished for my dream of a free Afghanistan for all Afghans. I fear that dream died in August of 2021 when the Taliban occupied Kabul.

But the American dream of freedom and equality is alive here. All I ask is to be allowed to live it. *Azadi rawest.* Free and legal. Was it worth it? you ask. Of course. Even a chance at the American dream always is.

ACKNOWLEDGMENTS

We have so many people to acknowledge that this section could be a book itself.

There's a Pashto proverb that says "Even on a mountain, there is still a road." We want to recognize a few of the people who helped us build a road through this mountain. First, twenty-five Marines were killed in action during our deployment to Sangin. These were the men who sacrificed their youth, hope, and potential. They were the centurions described by Marcus Flavinius in a letter to his cousin Tertullus in Rome. The centurion wrote:

> We had been told, on leaving our native soil, that we were going to defend the sacred rights conferred on us by so many of our citizens settled overseas, so many years of our presence, so many benefits brought by us to populations in need of our assistance and our civilization. We were able to verify that all this was true and, because it was true, we did not hesitate to shed our quota of blood, to sacrifice our youth and our hopes. We regretted nothing.
>
> But whereas we over here are inspired by this frame of mind, I am told that in Rome, factions and conspiracies are rife, that treachery flourishes, and that many people, in their uncertainty and confusion, lend a ready ear to the dire temptations of relinquishment. Make haste to reassure me, I beg you, and tell me that our fellow citizens understand us, support us, and protect us, as we ourselves are protecting the glory of the Empire. If it should be otherwise—if we should have to leave our bleached bones on these desert sands in vain—then beware the anger of the legions.

While we have the watch, we refuse to allow their sacrifice to have been in vain. We will say their names. Our lives will serve as testimony that we were witnesses to the greatest expression of love known to man. We will say their names:

Lance Corporal John T. Sparks
Corporal Justin J. Cain
Lance Corporal Phillip D. Vinnedge
Lance Corporal Joseph E. Rodewald
Lance Corporal Victor A. Dew
Lance Corporal Irvin M. Ceniceros
Lance Corporal Alec E. Catherwood
Lance Corporal Joseph C. Lopez
Lance Corporal James D. Boelk
Sergeant Ian M. Tawney
Lance Corporal Brandon W. Pearson
Lance Corporal Matthew J. Broehm
Lance Corporal Randy R. Braggs
First Lieutenant Robert M. Kelly
Lance Corporal James B. Stack
First Lieutenant William J. Donnelly IV
Sergeant Matthew T. Abbate
Lance Corporal Colton W. Rusk
Corporal Derek A. Wyatt
Sergeant Jason D. Peto
Lance Corporal Arden J. Buenagua
Lance Corporal Jose L. Maldonado
Corporal Tevan L. Nguyen
Lance Corporal Kenneth A. Corzine
Sergeant Jason G. Amores

There were men and women who, despite a Sisyphean task, held the line. In the middle of a spectacular failure, their valor in a sea of chaos never wavered. They gave their lives attempting to fulfill America's promise. We will say their names:

Lance Corporal David Espinoza
Sergeant Nicole Gee
Staff Sergeant Darin Taylor Hoover
Staff Sergeant Ryan Knauss
Corporal Hunter Lopez
Lance Corporal Rylee McCollum
Lance Corporal Dylan R. Merola
Lance Corporal Kareem Nikoui
Corporal Daegan William-Tyler Page
Sergeant Johanny Rosario
Corporal Humberto Sanchez
Corporal Jared Schmitz
Hospital Corpsman Third Class Max Soviak

There are 78,000 people that remain in Afghanistan who meet the SIV criteria. These men and women entered into a contractual agreement with the United States. They honored their part of the agreement and it's a stain on our nation's honor that they're abandoned and persecuted. We have a duty and obligation to continue advocating for these people who risked their lives in service to our nation.

It's May 4, 2022, as of this writing. Here are just a few of the messages Tom has received in the last twenty-four hours.

From a man named Homayoon:

Hello Dear Sir,
Hope you are fine and doing well. I really need your help and your support. I am in very bad condition because of the current situation in Afghanistan. I have worked with the U.S. military for 16 years. Please save my life.

From a man named Farid:

You left us alone in Afghanistan suffering.

And from Soltani:

Please help us. Save our life! You are the only hope for my family's future. My family needs help. I have children. Please save us! Do not leave us!

I'm forgotten and left behind in the country without being saved and evacuated after so many years where I was employed by the American government.

Since last August, Tom has received thousands of these messages. Tragically, so have tens of thousands of other veterans. We've become the face of deceit and we're bearing the burden of a lie. The collective guilt of our nation has fallen on our heads. We bear an invisible scarlet letter—Betrayer and Betrayed—one more moral injury to add to the collection.

Let's be the country these people believe we are. A country without honor will find it difficult to enlist young men and women to fight for it and even more challenging to find allies to fight alongside us.

At its core, our book is about family. While the book highlights two brothers and their struggle, we would not be here without our families. Our parents, siblings, and cousins made us who we are today. Our wives and children give us purpose, meaning, and hope for a better future. They are worth fighting for and, if necessary, dying for.

As men who are where we are because of teamwork and the contributions of others, we must thank the team at William Morrow/HarperCollins. Our editor, Matt Harper, was like a calm voice on the other end of the radio during a firefight. Day or night, through timely and steady editorial guidance, he got us to the objective and believed in our story enough to make this book happen. Lisa Sharkey was hugely supportive of this book from the start—believing in our story after following it in the media. Taryn Roeder in publicity saw something in our story worth sharing with other people and has done a great job helping us spread the word.

Worth Parker cold-called Tom in August 2021 to say he might have a way to get Zak aboard HKIA. A few weeks later we asked the retired Marine, Iraq and Afghanistan veteran, and writer to assist us in telling our story. Worth dedicated himself to putting readers on the ground we've walked for more than a decade. He remains committed to evacuating "just one more" Afghan who placed their faith in our nation.

Finally, while we are of different faiths, we both want to give all the glory to God the Father who is and was always faithful.

APPENDIX: "CONTACT FIRST"

"Contact First"
First Platoon, Kilo Company, Third Battalion, Fifth Marines
Support and Attachments
Sangin District, Helmand Province
September 2010–March 2011

Headquarters
Lieutenant Thomas Schueman
Staff Sergeant Timothy Henley
Sergeant Julio Mercado
Lance Corporal Eric Rose

First Squad
Sergeant Trey Humphrey
Corporal Matt Bland
Corporal Marcus Chischilly
Corporal Sean Leahy
Corporal Tevan Nguyen
Lance Corporal Darrell Cain
Lance Corporal Zach Evans
Lance Corporal Zach Hamilton
Lance Corporal Josh Hamlett
Lance Corporal Kevin Hinkley
Lance Corporal Nate Mori
Hospital Corpsman Third Class Doug Long

Second Squad
Sergeant Jeffrey Iwatsuru
Sergeant Fernando Sotelo
Corporal Raymond Orozco

Corporal Eric Sandoval
Lance Corporal Andre Gatlin
Lance Corporal Travis Haggerty
Lance Corporal Robert Murphy
Lance Corporal Maverick Voight
Lance Corporal Ryan Wishart
Hospital Corpsman Third Class Jay Bergstrom

Third Squad
Sergeant Jonathan Decker
Sergeant Brendan O'Bryant
Corporal Carlos Gonzales
Corporal Justin McLoud
Corporal Jared Smith
Lance Corporal Brenton Beever
Lance Corporal Tristen Chunn
Lance Corporal Michael Disanto
Lance Corporal Derek Goshke
Lance Corporal Justin Teague
Hospital Corpsman Third Class Rashad Collins

Support and Attachments
Master Sergeant Anthony Pierre
Staff Sergeant Robert Perez
Staff Sergeant Cassel Wiggins
Sergeant Matt Abbate
Sergeant Emilio Avalos
Sergeant John Browning
Sergeant Patrique Fearon
Sergeant Robert Gillespie
Sergeant Joe Myers
Sergeant Rane Rohweder
Corporal Jose Aguilar
Corporal Alex Harmon

Corporal Shon Hicks
Corporal Royce Hughie
Corporal Jordan Laird
Corporal Joseph Nikirk
Corporal FNU Roberts
Corporal Jacob Ruiz
Corporal Jeff Sibley
Corporal Michael Spivey
Lance Corporal Arden Buenagua
Lance Corporal Nick Calhoun
Lance Corporal Nico Detour
Lance Corporal Cameron Dunbar
Lance Corporal Alberto Espinoza
Lance Corporal Carlos Garcia
Lance Corporal Jacob Greene
Lance Corporal Clay Harrison
Lance Corporal Kevin Hensen
Lance Corporal Darin Hess
Lance Corporal Syrus Isfahani
Lance Corporal Jorge Ortiz
Lance Corporal Travis Smith
Lance Corporal Britt Sullivan
Lance Corporal Justin Talbert
Lance Corporal Bobby Thrillkill
Lance Corporal Colby Yazzie

GLOSSARY OF TERMS

ANA: Afghan National Army—The ground combat element of the Afghan National Security Forces

ANSF: Afghan National Security Forces—The Afghan military, police, and intelligence forces

CO: commanding officer—The senior officer in a military unit

COP: combat outpost—A small, well-fortified position from which military forces can plan and execute operations

FOB: Forward Operating Base—A secure, well-fortified position from which smaller positions such as combat outposts may be supported

HKIA: Hamid Karzai International Airport—The major international airport in Kabul, Afghanistan

IED: improvised explosive device—A "homemade" explosive device, typically crude in construction and made of military and/or nonmilitary components

IOC: Infantry Officer Course—The course by which the U.S. Marine Corps trains new infantry officers

ISAF: International Security Assistance Force—The umbrella headquarters over all allied (non-Afghan) forces in Afghanistan

JTAC: Joint Terminal Attack Controller—A service member trained and qualified to coordinate aircraft in providing close air support to ground combat forces

MEU: Marine Expeditionary Unit—An expeditionary Marine Air-Ground Task Force embarked aboard U.S. Navy amphibious ships. There are three MEUs deployed at any time.

MOS: Military Occupational Specialty—The specific job a Marine is trained to perform

OCS: Officer Candidates School—The process by which aspiring U.S. Marine officer candidates are evaluated for potential to serve as U.S. Marine lieutenants

PB: patrol base—A temporary position from which a squad or platoon conducts operations

PKM: A belt-fed, general-purpose Soviet bloc machine gun

RPG: rocket-propelled grenade—A portable, unguided, shoulder-fired launcher for explosive warheads

SIV: Special Immigrant Visa—The special program by which interpreters and people who worked directly for the U.S. government or companies in the direct employ of the U.S. government could receive a U.S. visa

TBS: The Basic School—The six-month school in which all Marine lieutenants are trained as provisional infantry before attending training in their specific MOS

XO: executive officer—The second-in-command of a military unit

A NOTE ON SOURCES

The authors wish to note the important contributions of two authors to the body of work about the American war in Afghanistan generally, and in this book specifically.

American author, Marine combat veteran, and former Assistant Secretary of Defense for International Security Affairs, Francis J. "Bing" West has written extensively on wars and warfare since 1972. Mr. West spent extensive amounts of time in Sangin District with Kilo Company, Third Battalion, Fifth Marines. Not satisfied to receive briefings in air-conditioned tents, he walked with Third Platoon, seeking the truth in every step. His efforts resulted in *One Million Steps: A Marine Platoon at War*, a book that refreshed our memories, validated our records, and clarified events obscured in the haze of combat memories.

CARTER MALKASIAN, PHD, is a professor, author, military historian, and national security practitioner with extensive time serving in, and advising the highest levels of American leadership about, Afghanistan. He served on a Provincial Reconstruction Team in Kunar Province in 2007. His Pashto fluency earned him accolades as a State Department political officer in Garmsir District of Helmand Province between 2009 and 2011. He returned to Afghanistan from May 2013 to August 2014, serving as political advisor to the Commander of the International Security Assistance Force, General Joseph Dunford. From 2015 to 2019, Dr. Malkasian served General Dunford as the Special Assistant for Strategy to the Chairman of the Joint Chiefs of Staff. His brilliant *The American War in Afghanistan: A History* offered clarity on dates and facts of strategic decisions deeply impacting both authors. As it concludes in 2020, the authors hope future editions will include Dr. Malkasian's astute analysis of the year that ended the dream of a free Afghanistan.

ABOUT THE AUTHORS

ZAINULLAH "ZAK" ZAKI is one of nine children and was raised by a subsistence farmer in eastern Afghanistan. Zak served as an interpreter for the U.S. forces during the war in Afghanistan with the Third Battalion, Fifth Marines in Helmand Province beginning in 2010, and later worked for the U.S. government in Kunar Province from 2012 to 2014. After more than six years battling bureaucracy with Major Tom Schueman as his advocate, Zak successfully immigrated to America with his family in 2021. He is currently awaiting a final ruling on his immigration status and working in construction in San Antonio, Texas. Zak is a husband and a father to his wife, Deewa, and their children, Subhanullah, Husna, Taqwa, and Sadaf.

MAJOR TOM SCHUEMAN served in Afghanistan for sixteen months, including the single bloodiest battle of the war in Afghanistan as a platoon commander with the Third Battalion, Fifth Marines in Helmand Province. Schueman redeployed to Afghanistan as a JTAC and advisor to the Afghan National Army while he was a member of First Reconnaissance Battalion. Schueman went on to get his master's in English literature at Georgetown University and teach English literature at the United States Naval Academy. He is currently a student at the Naval War College and remains on active duty. He is also the founder of the nonprofit Patrol Base Abbate. Tom is a husband and a father to his wife, Andrea, and their children, Amelia, Jack, and Marty.